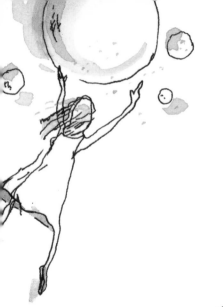

Vignette

Jane Lopes

Find more education, puzzle
solutions, and other fun stuff at
janelopes.com/vignette

Vignette

Stories of Life & Wine in 100 Bottles

Jane Lopes

with illustrations by Robin Cowcher

Hardie Grant
BOOKS

I always hated wine.

That is to say wine always made me feel socially uncomfortable. There's a kind of class system to it, especially around the fine stuff, and a perception that if you didn't come from a moneyed background, you couldn't possibly know how to appreciate it. Wine people use terms like "the bouquet of wet dog" to assert their superior knowledge over others less enlightened. Hearing talk like that, I shrank away and closed my mind to wine's possibilities, because if you grew up in back-country Taranaki, New Zealand, you knew three things:

#1   A "wet dog" is to be avoided at all costs, especially one that has been chasing sheep in a muddy creek.

#2   Grapes grow wild along the roadside and are only for foraging to eat as a treat. A basket press? That must be from the laundry.

#3   Wine is served from a cardboard box, except when – on very special occasions – Uncle Tracy turns up with the "good stuff": a bottle of Henkell Trocken. (Although I never tasted it, I was in awe of the gold foil and exotic-sounding German words.)

My first major interaction with wine came when I discovered the forgotten bottles left in the cellar of Uncle Tracy's new house. At age 13, having no interest in wine, I decided to make a little money by selling the dusty bottles to the older kids up the road. I did, however, help myself to some Coca-Cola, late 1950s vintage (an excellent year, showing buttery hints of barnyard caramel).

When I qualified as a cook at age 18, wine was something we served to customers; it definitely wasn't something for us lowly cooks – and it wasn't something any of us could afford either. As I said earlier, wine has its own class system: most restaurants survive off wine sales, but most cooks will never get to taste those wines.

Jane Lopes came into my life at just the right moment. A perfect storm of frustration had continued to build for the years that I'd been working as a chef and then running my own restaurant. Boy, did Jane have her work cut out with me – she must have thought I needed therapy with my wine issues! But when talking to Jane for the first time – she in NYC, me in Melbourne – I suddenly found myself becoming

6

passionate about wine. I didn't feel self-conscious. Jane and I agreed that the whole natural wine thing was a bit misleading, and we bonded over our opinion that bone-dry Aussie Rieslings can be pretty darn rough, preferring a little residual sugar in those wines (now who sounds like a wine snob?). I knew I had found the person I wanted to oversee the Attica drinks program: someone who could democratise wine but still have a creative vision based on our shared principles of quality and integrity; someone who was an absolute expert on wine, but who didn't act like one; someone who I knew would help people make great, informed decisions about what to drink.

But how to convince Jane to completely move her life halfway around the world? I'm certain that she wasn't persuaded by my wine knowledge. Nevertheless, Jane took the opportunity and ran with it, developing our strongest-ever drinks program and team, in all the ways that matter to guests and to a chef-owner like me.

The three things I now know about wine, thanks to Jane:

#1   Wine is mostly about farming. Humble winemakers like the great Rick Kinzbrunner of Giaconda will tell you, "All my work is in the vineyard." The last time I visited Rick he was fixing the vineyard motorbike. You won't read about that in the next issue of *Decanter*.

#2   The future of wine is women. It is not in men-only beef and Burgundy clubs. Men can't help but let their egos get in the way with wine. In addition to the amazing Jane, it was my better half, Kylie, who "forced" me to make the first visit to Giaconda. (She now calls Rick my "boyfriend.") It was women who got this man into wine.

#3   If you like a wine and it was made by people who care about the environment (and therefore care about you), it is good wine. Nothing more to say.

I'm certain by reading this book you'll learn to love wine – if you don't love it already – in the same way that I did: from Jane. Wet dogs and all.

– Ben Shewry

This book can be read in three distinct ways.

It is my hope that these three avenues braid themselves into a unique pattern for each reader, revealing an entirely new book to each person and on each sitting.

This book, at its most basic premise, delivers a list of wines (and a few beers and spirits) that are suggested drinking. Each chapter sets forth a style of wine (i.e. Champagne, Merlot, California Chardonnay) and then delivers one or more bottles that exemplify it.

Choosing the specific wines to represent each category presented its challenges. By no means are these choices designed to be "the best" or "my favorites." The bottles were chosen because they are among the most classic and important bottlings of their style, and/or because they have a personal connection to the chapter at hand. I've tried to choose wines that are relatively available and affordable, but when these parameters clash with the above, I have gone with classic and important over available and affordable. To remedy this practical limitation, I've included an appendix in the back of the book to offer five value suggestions for each style.

The other criteria that these wines must meet is a certain purity of intention. I hate to use the word "moral," as I think it's often misunderstood and misapplied to the world of wine, but I must agree with the winery's practices as I understand them. This does not mean choosing only small wineries, or only wineries who grow their own fruit, or only wineries that follow strict organic, biodynamic, or "natural" guidelines. What it does mean is choosing wineries that show consistent respect for the land they work, the grapes they grow, the people they employ, and the customers who are drinking their wine at the end of it all.

A perennial problem in the world of wine is how to list wines; there's no one agreed-upon way. For our purposes, the first part is easy: the producer. What comes next, in italics, is what we'll call the "cuvée name." This can be a few different things: a fantasy name generated by the producer (i.e. *Grande Cuvée, Grange, Quintet, Columella*), a vineyard site that is not its own appellation (i.e. *Kistler Vineyard*), a quality designation (i.e. *Kabinett, Federspiel, Reserve/ Reserva*), an age statement (i.e. *Blanco, 10 Year*), or a grape variety, if the wine is labeled by one (i.e. *Riesling, Pinot Noir*). What comes after the italics is the appellation and/or region. For this, and the cuvée name discussion, it's important to understand what an appellation is. An appellation is a legal designation for a wine region that often (particularly in Europe) also has regulations about what the wine is and how it's made. Champagne, Alsace, Sonoma Valley, Chevalier-Montrachet Grand Cru, and Verdicchio dei Castelli di Jesi Riserva are all appellations (yikes!). To keep things consistent, the

9

appellation will never be listed in italics, only as the first part of the larger region at the end of the listing. This will mean that some wines – like Stella di Campalto, Brunello di Montalcino (the appellation), Tuscany, Italy (the region) – do not list a cuvée name.

This is not a list of the 100 bottles to drink before you die, but rather a list of 100 bottles to live with, learn from, and find your truth in. It's the starting point, not the ending point. The list is designed to give a holistic view of the world of wine: if you were to drink each bottle on the list (or their suggested alternatives), you would have a nearly complete education on the flavors, weights, textures, and emotions that exist in a bottle of wine. At this level, the book acts as a trusty reference – a list that is always there when you are deciding what to drink next.

#2

The next layer of the book illuminates the emotions within and surrounding a bottle of wine. My adult life has been marked by moments of extreme joy, severe pain, wonder, accomplishment, failure, fear, solitude, community, and grit. Most times, a bottle of wine has been somehow involved: its emotions feeding mine, and my experiences providing a context for it. Each chapter not only proposes a style of wine and its exemplifying bottle, but also a vignette – a story from my life – that radiates the emotional truth of that wine. At this level, the book can be brought to the beach or on an airplane and read from cover to cover as a complete narrative.

#3

The final way to read this book is through the illustrative content within each chapter. Gone are the days (if ever they were here) when we just learned through reading words on a page. We think in images, colors, textures, music, games, puzzles, charts, graphs, maps, and surveys. This content is meant to present wine in a way that engages every level of our imagination. They are snippets – glances, nods – that bring a relevant aspect of each style to life. They are not meant to be exhaustively educational, nor are they meant to be irreverent fluff. These are entry points that allow the reader to gain a bit more understanding of the style of wine at hand. They present the novice with fun and relatable access to the style, and they present the expert with new ways to engage with familiar material. Some chapters present more challenging concepts than others, but all are designed to allow takeaways from multiple angles. At this level, the book is a perfect coffee table book: ripe with beautiful images that provide brief moments of information for the casual reader.

VIGNETTE

Included in these segments are *Somm Surveys*. To represent the broader sommelier community, I've enlisted some friends to share their opinions on topics relevant to certain chapters. There is a second appendix in the back of the book that lists all these sommeliers, their affiliations, and how to find them on social media.

One last (mini) layer is the glossary in the back. Any wine terms that don't quite get the explanation they deserve in the text are **bolded**. This means that they have a definition in the glossary, presented alphabetically.

—

Taken all together, these layers present a deservingly complex view of the world of wine. Sometimes, we just want someone to tell us what to drink (yes, even sommeliers want that). Sometimes, we want some more information about that style of wine, but without the dryness and pretension that often surrounds wine. And sometimes, we want to feel something about that wine. This book invites you in, at all three levels, to explore the magical and ever-growing world of wine and to unfold what it means to you.

Happy drinking,

j.

12

Champagne

A popular idiom among the sommelier set is that Krug tastes like failure.

Krug, one of the greatest Champagne houses of all time, actually tastes nothing like failure. It tastes like brioche and meyer lemon, honeysuckle and brûléed apple, white pepper and seashells. It balances opulent richness with laser-sharp acidity, allowing the wine to feel both abundant and fresh. Krug *Grande Cuvée* is a testament to the classic ideal of blending in Champagne; though many houses (including Krug) make wines from a single **vintage**, a single grape, or a single vineyard, Champagne has historically been about the art of the blend. The *Grande Cuvée* is a blend of all three major Champagne grapes – Pinot Noir, Pinot Meunier, and Chardonnay. It is a blend of over a hundred different sites throughout Champagne. And it is always a blend of over ten vintages.

Those who say it tastes like failure are candidates for the most difficult accreditation in the world: the Master Sommelier Diploma through the Court of Master Sommeliers. The Court of Master Sommeliers' website advertises that approximately 10% of candidates who sit for the theory portion pass it. That's a 10% pass rate on the first of three sections needed to become a Master Sommelier. (And this is after one has passed three different levels to even reach this point.) Those who pass the first section convene to take the other two sections – tasting and practical. Usually there are about

60 candidates. In 2017, eight passed. Sometimes it is less than five. Rarely is it over ten. At the reception to publicly announce and celebrate the new Master Sommeliers, Krug is served. Most people there have just failed one or more parts of their Master Sommelier examination.

The test is not for everyone. Most of us agree there are hundreds of masterful sommeliers who do not engage with the Master Sommelier exam. But for those it suits, it not only builds knowledge and skill – it creates community and character like nothing else.

CHAMPAGNE

The Origin Story of the Bubble

The Big Bang:
Champagne Method

The most traditional and highly esteemed method of making sparkling wine is appropriately dubbed the Champagne method. This is where a second **fermentation** takes place in the bottle: carbon dioxide, a natural product of fermentation, is trapped in the bottle and turns into fine bubbles. This is a meticulous process that results in the most elegant and precise sparkling wine in the world.

A Happy Accident:
Ancestral Method

In the primitive days of winemaking, fermentation would often cease in winter due to cold temperatures. Unsuspecting vignerons would bottle these wines, still with residual sugar, and when the ambient temperatures rose and the yeast activity reignited, fermentation would start again. Carbon dioxide, this time restrained by the pressurized bottle, would stick around as bubbles rather than disappearing into the ether. The main difference here versus the Champagne method is that these wines only go through a single, halted fermentation, rather than two separate ones. As a result, many are lower in alcohol and effervescence, and have some sweetness.

Tropical Paradise Meets Tank Living:
Charmat Method

Prosecco is the most famous utilizer of this process: wines undergo secondary fermentation in a pressurized tank, instead of in a bottle. A bit of the romance is gone, and the **lees** exposure that defines the aromatics and taste of Champagne is notably absent. The resulting flavor profile of these Charmat-made wines is fruity, bright and primary. They may not be quite as complex as their Champenoise brethren, but they can be well crafted and pleasing nonetheless.

Industrial and Lifeless Effervescence:
Injection Method

No one is out there espousing the qualitative merits of injecting CO_2 into wine to make it sparkling. This is the same process used to make soft drinks, and results in large, harsh bubbles that dissipate quickly in the glass. It is low cost, however, so if the goal is a bottle of sparkling wine around the $5 mark, this is the place to be.

California Chardonnay

16

When I was growing up, my parents finished most evenings with a nightcap.

Scotch and soda for my dad, gin and tonic for my mom. I learned how to make these drinks at an early age. My sister and I were happy to do so: a gesture towards growing up and an opportunity to do something nice for our parents.

When my parents did drink wine, it was California Chardonnay. Oaky and cheap, these wines stunk of canned pineapple, toffee pudding, and buttered popcorn, with a burn on the finish that reminded me of nail-polish remover.

I rebelled. My first – and to this day, great – loves in wine are the searingly tart Rieslings of Germany and the austere Nebbiolos of northern Italy. Wines that reject the flabbiness and oak that my parents' beloved Chardonnays relied on so heavily. (An **MS** once told me that Nebbiolo is for people who like to sleep on nails, play rugby, and eat kale. I don't disagree with the sentiment – of the grape's punishing character – but I contest the negative valuation.)

When I started working in the wine industry, my parents and I developed a vocabulary for what they liked about these wines. They did like the flavor of oakiness but, more to the point, they didn't like citrus-driven or acidic wines. I introduced them to Rhône whites, Friulano, Pinot Gris, white Priorat, and great California Chardonnay. They took a chance on the not-always-oaky and the never-cheap. And I took a chance on not punishing myself. The round, ripe, silky mouthfeel of these wines was comforting to them, and as I sought out these wines for my parents, they became comforting to me.

17

CALIFORNIA CHARDONNAY

Oakiness vs. Price in California Chardonnay

This isn't meant to be an exhaustive list of every California Chardonnay. Some great ones are indeed omitted. It is meant to show the correlation between the two factors most often discussed on a restaurant floor and some deviations from that pattern.

Price

Price based on average or stated retail price in US$

Impression of Oak based on an algorithm of [time spent in oak] x [percentage of new oak], with some leeway for interpretation based on my own impression

You can't justify this

$250 ● Marcassin

$200 ● Aubert

● Peter Michael

$150 ● Kistler   ● Kongsgaard

● Littorai

● Williams Selyem

● Hyde de Villaine   ● Staglin

● Mount Eden   ● Ramey

$100 ● Hanzell   ● Patz + Hall   ● Far Niente

● Rochioli   ● Newton

● Chateau Montelena   ● Calera   ● Grgich Hills

● Stony Hill   ● Cakebread

● Mayacamas

$50 ● Sandhi

● Au Bon Climat   ● Sonoma-Cutrer   ● La Crema

● Rombauer

● Four Vines   ● Kendall Jackson

You can't afford to make this

Impression of Oak

Rum

In one of their more brilliant parenting moves, my mom and dad encouraged me and my sister to try alcohol early on in life.

A taste of a Corona at six. A sip of the dreaded Chardonnay at nine. A lick of Scotch at eleven. This was enough to make any sane child swear off alcohol forever. *Who would willingly drink these things?*

Although my sister had similar responses at the time, like most teenagers, Beth soon discovered the pleasures of alcohol. As for me: I still couldn't stomach the taste of it, and I soon learned that I didn't like how it made me feel. I had grown into a fierce little control freak, and alcohol made me feel fuzzy and soft. My body and mind rebelled against this lack of sharpness, a tension that created my first taste of anxiety.

As socializing, meeting boys, and making friends often centered around drinking in high school and college, I began to feel I was being left behind. My first year in college was an epic failure. I had anticipated it being my salvation, but instead it was another metric by which I felt my social ineptitude. I rarely went to parties. When I did, instead of the revelrous, booze-fueled camaraderie felt by most undergrads, I just felt my difference sting more sharply. By the time I finished that first year, I had never been on a date, never even kissed a boy. I transferred schools, thinking maybe these just weren't my people.

The summer in-between, I made plans to go to the Dominican Republic. This was a four-week trip, part community service, part adventure, with other undergrads from around the country. *These are going to be my people!* I thought.

Immediately, I knew I was wrong. The girls looked like women. They wore skimpy bathing suits and talked about sex and had figured out hair removal. I wore soccer shorts and read Gabriel García Márquez and had not really figured out anything. I made one friend, Dave, who was loud but gentle. He shared his favorite music and taught me how to play Euchre.

One sweltering night, underneath the thatched roof of our open-air hut, Dave called to me. I was reading a book on a corner bench, looking down over the hills leading to Cabarete.

"Hey, Jane, you ever play flip cup?"

"Nope…" I said, the hesitation clear in my voice.

"Come on, it'll be fun!"

Flip cup is a ridiculous game, when you think about it. A generation of college students have forgotten everything they knew about Foucault and Freud, but have cultivated the unique ability to flip a plastic cup 180°.

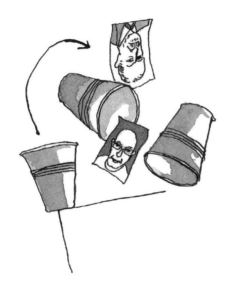

I was surprisingly good at it. I assumed that being good at flip cup meant that you could drink less beer, but I was mistaken. The beer was hard to get down, but after a while I began to feel a warm, fuzzy haze that was not altogether unappealing.

The good vibes shifted later that night as my stomach turned sour while I slept. I crawled out of my bed in the lofted room the girls shared and found my way to the 20-foot staircase that was our only way out. Each step was about six inches long, not nearly enough runway for my size-11 feet. I backed down the steps, sliding off nearly every single one, gripping the handrails with each new bout of nausea. I finally made it to the ground and raced around the corner to the communal bathroom just in time. I threw up what felt like dozens of times before passing out in exhaustion next to the bowl. Our house mother found me in the early hours of the morning and helped me back up the stairs to bed. Her feet seemed to have no problem fitting elegantly on each step.

I didn't play flip cup again that trip (or ever again, come to think of it), and the only booze I drank was small sips of rum here or there. The rum was oddly soothing. Sweet, but with no real sugar, strong but not harsh. It seemed to coat my throat and calm my stomach, and was nearly cheaper than water on the island.

I would return to the Dominican Republic over ten years later, after my first attempt at the Master Sommelier exam; I hadn't figured out too much more by then.

VIGNETTE

A Style Map of Rum

Rum production in the Caribbean and Central/South America can be divided into three main categories: Spanish, English, and French. The following map shows the major countries of rum production and the main producers in each. Sugar-product distillation meets geography, history, language, and politics.

Spanish-speaking countries and islands generally produce a lighter, rounder and more rectified flavor profile. Spanish rum is often referred to as "ron."

English rum is historically from British colonies, who would ship it back to England for aging and bottling. It is generally fuller, darker and more aged, with greater molasses character.

Unlike Spanish and English rums, which are made from molasses, French rum is distilled from fermented fresh sugarcane juice. Referred to as "rhum" or "rhum agricole," it is column-distilled, and dry and vegetal in style.

Bermuda
Gosling's

Cuba
Havana Club

Dominican Republic
Brugal, Barceló,
Atlantico, Matusalem

Puerto Rico
Bacardí, Don Q

St. Croix
Cruzan

Jamaica
Wray + Nephew,
Blackwell, Appleton,
Myers's

Haiti
Rhum Barbancourt

Martinique
Rhum JM, Neisson,
Clément, La Favorite

Barbados
Mount Gay

Guatemala
Zacapa,
Ron Botran

Nicaragua
Flor de Caña

Panama
Ron Abuelo

Venezuela
Diplomático, Pampero,
Santa Teresa

Trinidad + Tobago
Angostura

Colombia
Dictador

Guyana
El Dorado

Vino Italiano

My dad turned me onto the Polish author Jerzy Kosiński.

Why a man would give his 18-year-old daughter a copy of *The Painted Bird* is beyond me, but somehow the chronicle of horror fascinated rather than repulsed me. Kosiński defied death in World War II Poland, then again when he was supposed to be at Sharon Tate's house the night of the Manson murders but had to stay in New York because of a lost piece of luggage.

It was the fall of 2005, and I was reading Kosiński's *Steps* on a plane from California to Rome. It was my junior year of college, and the third year in a row that I would kick off knowing no one. I transferred my sophomore year to the University of Chicago. These were much more my people, but I was dreading having to start over again with a new set of students on my quarter abroad in Italy. Kosiński writes: "Had it been possible for me to fix the plane permanently in the sky, to defy the winds and clouds and all the forces pushing it upward and pulling it earthward, I would have willingly done so. I would have stayed in my seat with my eyes closed…and I would have remained there, timeless, unmeasured, unjudged, bothering no one, suspended forever between my past and my future."

I felt exactly the same way.

—

The first nights in Rome were hard. It was a bit more of a party crowd than I was used to, and I felt that familiar sting of not quite fitting in. There was a phone booth in our hotel, and I spent those nights crouched on its floor, sobbing to my parents. I thought I knew why I was crying, but the tears also seemed to have a life of their own and came out without much prodding.

At our first big dinner, all 20 of us sat outside on a cobblestoned street just off a big piazza. There were carafes of wine on the table: no named grapes, no labels, just a red one and a white one. Wine was part of the meal in Italy, and with each sip I felt more and more at ease. Strangely, it wasn't about being intoxicated. It was about feeling connected to the food I was eating, the country I was in, and the people at the table.

Rome ended up being somewhat revelatory for me. That can't-hold-back-the-tears feeling happened less and less. I was less self-conscious about my height and my body. I felt more adventurous, braver, more confident, and more valuable than I ever had before.

These emotions solidified one eventful night in Capri. A group of about ten of us had planned a weekend away on the island. The main attraction was the Blue Grotto – a cave off the coast that, due to magical reflections beyond my understanding, shone blue at certain points in the day. When we arrived on the island, we were told by our hostel mom that there weren't any tours of the grotto that day due to high tide and jellyfish. She added, with a gleam in her eye, "You could probably sneak down there and jump in, if you wanted…"

We went down and tip-toed up to the edge of the cliff. The water was rough and (I imagined) teeming with jellyfish. We stripped down to our bathing suits, though I'm not sure if any of us believed at that point that we would go through with it. But someone was the first, then another followed. I knew at that moment that it would be more painful *not* to jump in than the pain any jellyfish could cause. I knew then that I would rather deal with the consequences of taking a risk, rather than the regret of not ever having gone for it.

So I jumped. I hit the water and immediately felt exhilarated. Alive. I don't think I would have even felt a jellyfish sting me; my body was too busy radiating joy and vivacity. I swam to the entrance of the grotto and saw the magnificent blue light bouncing off every corner. I was outside of myself – outside of the insecurities, outside of the anxiety, outside of the pain. Inside life.

That night, we had a casual dinner with carafes of wine on the table. Somehow a bottle of Brunello di Montalcino made its way into our possession, and I had my first real taste of fine wine. We drank limoncello until the sun came up, then passed out on the lounge chairs of our giant patio overlooking the ocean. I woke up nuzzled in the arms of a boy who would become my first real kiss.

I thanked Italy – its historical majesty, its natural beauty, and its zest for life and wine – for ushering in a new era of my life. I felt like myself on that trip, an identity that I would veer closer to, and further from, in the years to follow.

24

Wine Whims That Became History

A great many decisions in the history of wine can be credited to the sometimes-ludicrous whims of religious, royal, and political leaders. These men (let's be honest) knew that their word was law and, as a result, thought that every idea was an epiphany. Some of these practices and traditions persist today.

Prince Alberico Boncompagni Ludovisi
1946

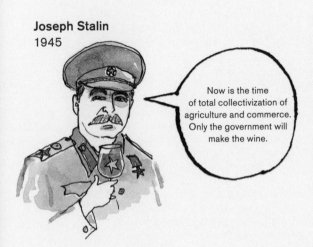

I inherited my family's estate, Fiorano, just outside of Rome, and introduced the international grapes Cabernet Sauvignon, Merlot, and Sémillon.

Outcome: The wines of Fiorano were not discovered by the world until the 1970s, when famed food writer Luigi Veronelli began telling the prince's story. These wines never became more than a niche, as the prince pulled out nearly all his vines following the 1995 vintage. But his impulse to plant international grapes contributes to an important narrative in Italian wine, with the emergence of the "Super Tuscan" in the 1970s. There may have even been direct causality from the prince to the Tuscans, as in 1966 the prince's sole daughter married Piero Antinori, who would go on to create the famed *Tignanello*.

Joseph Stalin
1945

Now is the time of total collectivization of agriculture and commerce. Only the government will make the wine.

Outcome: The progress of wine-growing regions behind the Iron Curtain screeched to a halt. Tokaj, a legendary region for sweet wine in northeastern Hungary, saw most of its vineyards commuted to vine training that allowed for mass production. Only when the curtain fell in 1989 did the region revert to a quality-minded approach, with new ventures and international interest blossoming in the subsequent 30 years. Many other countries – Bulgaria, Georgia, Romania, the Czech Republic – are still picking up the pieces.

Pope Clement V
1309

I don't want to live in Rome. We're going to move the papacy to Avignon.

Outcome: The 67 years that the papacy was in Avignon brought money, development, and prestige to the Southern Rhône, and specifically the region of Châteauneuf-du-Pape. Pope John XXII succeeded Clement and worked hard to improve and promote these wines. He erected the castle that is now symbolic of the region. Today, Châteauneuf-du-Pape is considered one of the great wine regions of France.

Charlemagne
800

This red wine can be messy and my wife doesn't like the way it stains my beard. I have no intention of giving up wine, so let's plant white grapes instead!

Outcome: Corton-Charlemagne, one of the great **Grand Crus** in Burgundy, was replanted with white grapes during Charlemagne's rule (probably Pinot Gris, Pinot Blanc, and Aligoté). Post-**phylloxera**, Chardonnay became the grape of favor in Burgundy, but the tradition of white wine stuck. Today, Corton-Charlemagne is the source of some of the greatest white wine in the world.

26

FILIPA PATO *FP BRANCO* Bairrada, Portugal
HERDADE DO ESPORÃO *TINTO RESERVA* Alentejo, Portugal

Corked Wine

or PORTUGUESE WINE

I drank quite a bit of wine through the rest
of my college career.

I adopted the Italian approach: a bottle of white and red were always
on the table, but I wasn't too fussed about the graped or the region.
I graduated with an eye towards graduate school, but instead of writing
applications during my senior year, I planned to take a year off after
college to apply. I was hoping to work for the admissions department
at the University of Chicago, but I didn't get the job. Pondering what
to do for the year – my English literature degree granting me almost
no actual qualifications – I found a job posting to work in a wine store.

I bought *Wine for Dummies* and spent the weekend before my
interview reading it. Luckily, the manager Rachel was also a graduate
of the University of Chicago, so we spent more time talking about
Descartes and how ugly the Max Palevsky dorms were than wine.
She hired me.

My first day at LUSH was a Monday, in which the managers
spent the entire day tasting with wine distributors to suss out new
bottles for the store. I tasted along with them, trying to take note of
the different grapes, regions, and flavors that were flying across my
palate. But for the most part, I couldn't tell one from the next. My
mouth, tongue, and teeth hurt by the end of the day. My gums were
crusted with red goo, and I feared my teeth would never regain their
once shiny-white color.

At our last appointment, the owner of the store asked for my opinion on a wine.

"Ummm…I guess it's pretty nice…crisp. Dry." I couldn't taste a thing, let alone come up with words to describe it. It was a Portuguese wine with an exotic-sounding grape name that I had never heard of. I was hoping I came somewhere near the mark.

"It's corked!" he yelled, somewhat amused, somewhat annoyed.

"Well, yes, I…" I was about to explain that I knew it was corked, I could see the cork on the table, but it was obviously uncorked because we were drinking it.

Rachel blessedly cut me off. "Mitch, it's day one. Give her a break."

Rachel later explained what it meant for a wine to be corked. I expressed my gratitude to her for not letting me stick my foot in my mouth any further, and I promised that I would learn. I would not let her down.

Hotspots for Fresh Takes on Portuguese Wine

Portugal is best known for the **fortified** wines of Porto and Madeira. There are many classic styles of dry wine as well: structured and oaked Douro reds, the regal Touriga Nacionals of the Dão, and the spritzy blends of Vinho Verde. This map showcases wines that have not (yet) made it into the mainstream that offer exceptional value and unique expression.

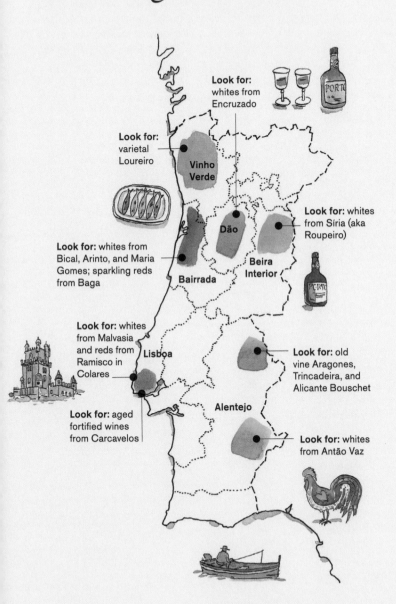

Look for: whites from Encruzado

Look for: varietal Loureiro

Vinho Verde

Look for: whites from Síria (aka Roupeiro)

Dão

Look for: whites from Bical, Arinto, and Maria Gomes; sparkling reds from Baga

Bairrada

Beira Interior

Look for: whites from Malvasia and reds from Ramisco in Colares

Lisboa

Look for: old vine Aragones, Trincadeira, and Alicante Bouschet

Look for: aged fortified wines from Carcavelos

Alentejo

Look for: whites from Antão Vaz

CORKED WINE

The Quick and Dirty on Wine Flaws

Below is a guide to wine faults: what they are, how to detect them, and what you can do to prevent them (or avoid coming across them). There are other issues in wine – like **volatile acidity**, **Brettanomyces**, and **reduction** – that can be flaws at high levels, but can also be desired characteristics (head to the glossary to find out more about these).

	CORK-TAINT (TCA)	OXIDATION
What it is	A compound in natural cork that can infest the wine itself or, with less frequency, take hold in cellars (so even wines not sealed under cork can be susceptible)	Oxygen exposure
What it smells/tastes like	Mold, wet dog, wet cardboard; the palate can feel hot (in terms of alcohol) and stripped of flavor	Dried fruits, cider, vinegar
How to prevent or avoid it	Buy wines not sealed under cork, or from wineries that purchase very high-quality cork. Learn how to detect cork-taint so you can send corked bottles back in restaurants and return them to retailers.	Oxidation can occur naturally (or prematurely) as wine ages. Storing wine on its side keeps the cork moist and prevents oxidation. Wine also begins to oxidize as soon as it is opened, so invest in a preservation system if you like to enjoy a bottle over the course of a week.

VIGNETTE

MADERIZATION	LIGHT-STRIKE	REFERMENTATION	MOUSINESS
Heat exposure	Blue and ultraviolet lights transforming amino acids into sulfides (and other less desirable compounds)	When yeast activity causes wines to start fermenting again, in bottle	A lactic bacteria taint that is not perceivable on the nose but can be quite prominent on the palate and retronasally
Caramelized fruit, nuttiness (can also be detected by a cork that is protruding or retreating)	Cabbage, sewage, musty newspaper	Wines get fizzy and cloudy	Natural wine advocate Alice Feiring has offered the descriptors "puppy breath" and "dog halitosis"
Don't store your wine above the fridge, or leave it in your car or next to a heat source.	Store your wine in a dark place, and buy wine in darker glass bottles.	There's not much as a consumer you can do to predict refermentation – take the bottle back to your retailer or send it back in a restaurant.	Low SO_2 levels and poor hygiene in the winery can lead to mousiness. Make sure you know and trust the winery if it purports to be of the "natural" movement.

CORKED WINE

Scotch

32

I've always maintained that if I can end each night with a glass of Scotch and fall asleep peacefully, everything will be fine.

Drinking has rarely been a tool I use to mask and forget when times are bad; it's mainly been a symbol of prosperity, of hope, of happiness, and – above all – of wellness.

I have been unwell for almost half my life. I can't pinpoint the exact day or time that things changed. There was no single moment, or if there was, I don't remember it. Just a sense, arising slowly and distantly, that I was not right. When it first started, perhaps my junior or senior year in college, I would describe it as a shift in my ability to focus. You know that feeling when you're drunk and you could just stare at a doorknob for ten minutes? And tearing your eyes from it requires supreme effort? That's how I started to feel all the time.

At first, it wasn't that disruptive. I felt a little off, but I was still happy. It had not yet become pervasive. Around the same time, I developed stomach problems which, in my 20-year-old world, were much more disturbing. My gut burned, I was bloated, and I rarely felt hungry. I lost weight, but my inflamed and protruding belly never let me feel thin. My stomach would feel better after I ran, so I took to the gym for four, five, six miles a day. I lost more weight. I balanced a heavy course load with acting in school plays, internships, and my running. My sleep suffered, but I was excelling in a very competitive academic environment and enjoying it. After an awkward and static high school career, and a belly flop of a first year of college, I finally felt dynamic, successful, and appreciated. But my success also felt tenuous. I wasn't just running at the gym. I was on a constant treadmill, afraid of the face-plant that would come if I let up.

By the time I graduated college, I had undergone a slew of tests, both for my stomach and my general malaise. Anyone who has had nebulous stomach problems understands the frustration and lack of answers in this arena. I was told I have IBS, which really means "your stomach hurts and we can't tell you why or help make it better." 33

I saw a doctor who told me I had an allergy to corn, dairy, nuts, citrus, many vegetables, every source of fat, and almost all grains. This, in my burgeoning wine and food career, was the worst fate imaginable. I couldn't eat anything my friends and family did, and the idea of going out to eat and listing all those allergies was inconceivable. I remember being on a solo trip to Boulder and sitting at the bar at Frasca, a mecca of food and wine, and scanning the menu. My eyes filled with tears as I realized: I can't eat anything here. I ordered a glass of wine and left. (Mercifully, I was not diagnosed as being allergic to grapes, and I took advantage of this.)

This was the beginning of being afraid of my body. I constantly thought about what I ate and drank, and lived in fear of how it would affect my health. After months on this restrictive diet, I felt worse and gave it up. Thus began a long tradition of spending time and money on purported solutions that only made me more crazed.

Whether it was the chicken or the egg, my mental health also suffered at this time. I never slept through the night. I would go to bed at 3am or 5am and wake up at 8am, exhausted but unable to sleep, my heart beating in my chest and ringing in my ears. I went the healthy route first: I tried giving up alcohol, eating a healthy diet, getting plenty of exercise, going to acupuncture, therapy, making time to relax. I tried a slew of different psychiatric drugs. Nothing worked. So, I went the unhealthy route. If I wasn't going to feel good, no matter how well I took care of myself, why bother?

I made bad decisions. I stayed out too late. I drank way too much. I tried recreational drugs.

When I couldn't sleep, I would slug down two or three shots of Scotch to help me get back to bed. Scotch seemed like the regal choice at that hour. I somehow felt romantic and literary when I drank Scotch at 3am, rather than desperate and frantic.

When I did these things to myself, I knew why I felt bad the next day. When I had taken care of myself, and still felt awful, there was no reason. This not-knowing was the truly maddening part. So I clung to the brief moments of pleasure my self-destructive behavior provided, and the false security that at least I knew why I felt so horrible.

One particularly bleak manifestation of this philosophy had me waking up in bed in my full outfit from the night before – coat, hat, shoes, bag still slung around my shoulder – covered in my own bright-red vomit. I had been asleep for over ten hours and was late for work. My memory of the night before was completely wiped out. I scanned my phone. At 8pm, I had called a cab. At 1am, I had sent a text to a boy I was seeing that read *Help, I'm scared*. No recollection of any of it. I sobbed in the shower as I washed the crime scene off my face and hair.

Small memories started to come back to me, and along with some recollections from others, I put the night back together. My friend Bridget and I had had an early dinner at a restaurant where I knew one of the chefs. I had called her a cab at 8pm. I wanted to stay out, so I sat at the bar and chatted with the bartender until my friend was done in the kitchen. We went across the street to a neighborhood bar and drank Scotch neat. That's where things got really cloudy.

Normally, when I drank heavily, I would have a hard time sleeping, I would never throw up, and I would remember the night with near-perfect clarity. Everything was different with this night, and after a little research, it seemed very much in line with having been slipped a date-rape drug of some kind.

When I told my chef friend, his immediate response was that "no one at that bar would do that." He expressed no concern for my wellbeing, just a misplaced irritation that I was somehow insulting his favorite bar. I felt guilty and ashamed. I felt like I had done something wrong. I thought that maybe I had just gotten too drunk, though in hindsight I know this was not the case. I swept it all under the rug, and never told my friend how much his reaction not just hurt, but really damaged me.

When I told Beth about being drugged, she flipped out.

"You could have been raped! Killed!"

"Yes, I'm aware."

"What were you thinking?!"

I know now that this was my sister being scared to death for me, loving me so much that she didn't know how else to react. But at the time, I felt attacked. She blamed me and my "lifestyle" for the incident. She thought I took risks and surrounded myself with untrustworthy people. If I hadn't been out late with people I didn't know, drinking at a random bar, none of this would have happened. But, I reasoned, should you not cross the street because you could be hit by a car? Our definitions of reasonable risk were incompatible. Mainly because I didn't fear the consequences of the risks I took.

35

The less fear I had for the consequences of my actions, though, the more fear I had of my body. I clenched every part of myself tight. It felt like if I didn't, everything might fly away.

If You Like This,
You Should Try...

There are many charts, graphs, and wheels to describe the complex flavor profiles of Scotch whisky. The below starts with the commonly known Single Malts profiled opposite and compares them to lesser-known distilleries by removing or adding a certain part of the profile.

Glenlivet

+ honeyed − floral = <u>Aberlour</u>
+ medicinal − nutty = <u>Arran</u>
+ fruity − vinous = <u>BenRiach</u>
− vinous = <u>Cardhu</u>

Macallan

− weighty − vinous = <u>Auchentoshan</u>
+ floral − nutty = <u>Balvenie</u>
+ floral − vinous = <u>Ben Nevis</u>
+ floral + peaty = <u>Mortlach</u>

Springbank

+ fruity − nutty = <u>Bowmore</u>
+ fruity + malty = <u>Bruichladdich</u>
− honeyed − vinous = <u>Oban</u>
+ fruity + peaty = <u>Highland Park</u>

Laphroaig

+ fruity + spicy = <u>Clynelish</u>
− peaty − weighty = <u>Isle of Jura</u>
+ fruity + spicy = <u>Talisker</u>
+ fruity + floral = <u>Old Pulteney</u>

Candied

Honeyed

Nutty

Malty

Weighty

Floral

Fruity

Spicy

Peaty

Vinous

Feinty

Medicinal

SCOTCH

38

White Burgundy

I was invited to a private Burgundy tasting the year I started working in wine.

My friend Reynold worked as a consultant, and his boss often held wine tastings for the staff. Reynold told me that we had to get there early because his boss would always open the best stuff at the beginning to encourage people to be on time.

We arrived at a beautiful penthouse apartment in the Gold Coast neighborhood of Chicago. We were up above the trees, with the exact right vantage point to see, seemingly before our eyes, the fall leaves changing from green to red to brown.

This was the perfect setting to be introduced to white Burgundy. Chardonnay from the Burgundy region of France is a category of wine that has mystified and entranced for centuries. These wines are always regal, always distinguished. The spectrum ranges from taut and laser-like to opulent and exotic; but it's the difference between Audrey Hepburn and Sofia Loren, not the difference between Audrey Hepburn and Janis Joplin.

White Burgundy's life cycle also seems to follow this model. White Burgundy believes in living a long, illustrious life, but it also believes in dying young. There is no in-between. White Burgundy does not want you to see it deteriorate or decay. I've never had a bottle of white Burgundy that I thought was "just over the hill" or "on its way out." They are either alive or dead.

The first bottle we tasted that night was very much alive, even at nearly 30 years old. It was 1983 Louis Latour Chevalier Montrachet. Louis Latour is a classic large-scale Burgundy **negociant** and, as such, is often maligned (or just ignored) by the sommelier community. These negociants, though, usually have the resources, expertise, access, and tradition to make some of the greatest wines in Burgundy. (I seem to be the only one in on this little secret, which is fine, as it keeps the prices out of the astronomical realm that Burgundy can all too easily reach.) These producers have proven to be overwhelmingly less susceptible to trends of the day, and therefore present pure and true representations of the land and vintage.

I was in utter disbelief that the flavors of this wine could be created simply through fermented grapes, aged in barrel and bottle. The wine smelled like milk chocolate, jasmine, almond oil, and quince. The palate was silky and plush, with still-lively acidity. It was like eating lemon pearls folded into crème brulée. Sommeliers often talk about having an "aha!" moment with wine; this was mine.

Reduction vs. Richness in White Burgundy

The graph on the opposite page is meant to help navigate the minefield of white Burgundy, dissecting two of its more important flavor parameters. These are complex concepts, so hold on!

The x-axis is a scale of how *reduced* the wine presents. This is a hot topic in Burgundy today, and many producers are evolving toward this style. To be clear, this isn't meant to be a comment on winemaking, but on how a wine presents in its finished state. Most Chardonnay in Burgundy today is initially handled *oxidatively*, meaning that grapes endure pre-fermentation oxygen exposure to provide resistance to oxidation at a later stage. A series of factors – from soil composition, to fermentation and aging temperature, to **lees handling** and **racking** – can then determine if a wine presents as reduced. The tasting notes for this characteristic include flint, gun smoke, and struck match. On the other end of the spectrum, the complete absence of these notes marks the oxidative style. These wines are not *oxidized*. They just lack the telltale smokiness of **reduction**, and instead present more primary flavors of fruit, oak, lees, minerality, or whatever else that wine is about. It is not a matter of better or worse, just personal preference.

The y-axis is a scale of richness. *Lean* white Burgundies might be harvested early, probably don't rely on a huge oak presence, keep minimal alcohols, and have piercing acidity. *Opulent* styles might gain their richness from fruit, oak, lees, or alcohol.

The focus here is on the white wine producers of the **Côte de Beaune**, with only a couple from Chablis and the **Côte de Nuits** included. Disregard who buys their grapes and who grows their own (many of these producers actually source in both ways, often through a second label), and instead focus on the flavor profile you like. Most of these producers offer everything from a humble Bourgogne Blanc all the way up to **Grand Cru** bottlings, allowing you to dip your toe in at whatever price-point is comfortable.

As this graph is based on tasting impressions rather than concrete facts, the placements are subjective and certainly up for respectful argument.

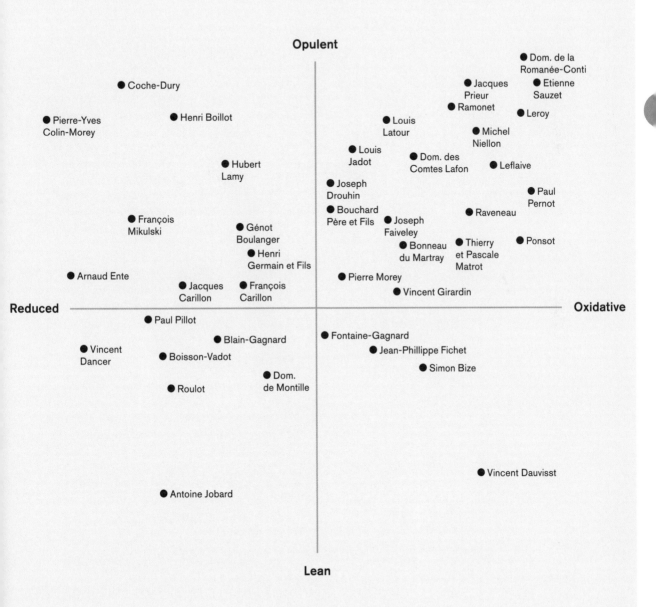

Opulent

Dom. de la Romanée-Conti

Coche-Dury

Jacques Prieur

Etienne Sauzet

Ramonet

Pierre-Yves Colin-Morey

Henri Boillot

Leroy

Louis Latour

Michel Niellon

Louis Jadot

Dom. des Comtes Lafon

Leflaive

Hubert Lamy

Joseph Drouhin

Paul Pernot

Bouchard Père et Fils

François Mikulski

Génot Boulanger

Joseph Faiveley

Raveneau

Henri Germain et Fils

Bonneau du Martray

Thierry et Pascale Matrot

Ponsot

Arnaud Ente

Jacques Carillon

François Carillon

Pierre Morey

Reduced

Vincent Girardin

Oxidative

Paul Pillot

Fontaine-Gagnard

Blain-Gagnard

Jean-Phillippe Fichet

Vincent Dancer

Boisson-Vadot

Simon Bize

Dom. de Montille

Roulot

Vincent Dauvisst

Antoine Jobard

Lean

WHITE BURGUNDY

Somm Survey: What was your "aha!" wine?

"1983 Château Rieussec. I'd bought the wine on auction when I was studying engineering; blew my mind that a beverage could invoke that reaction. It's probably the reason I have more Sauternes than I need in my cellar."
Benjamin Hasko

"1982 Haut Brion as part of the 'reserve pairings' pouring alongside the beef course at Charlie Trotter's for my 22nd birthday in 2007."
Caleb Ganzer

"After a life spent doubting that any of these verbose wine vocabularies actually existed in a glass of wine, Condrieu proved me wrong."
Carla Rza Betts

"1993 Champy Clos St. Denis. It was like hearing music for the first time. I learned more about wine culture and Aloha than Pinot Noir, and that one tasting has shaped the way I treat others who are beginning to learn."
Chris Ramelb

"I was working at an Italian restaurant in downtown Denver, and there was a wine bar I would visit after work that had 1995 Château Margaux for $250 (!). I would think about it all the time, but in 2000 that was serious money for a 23-year-old. On the last day of business, we had a big party and I made some significant tips and decided that was the day. I sat at the bar by myself and enjoyed that wine like few since."
Eric Hastings

"2001 Château Rayas. I didn't know wine could smell and taste like white pepper."
Jackson Rohrbaugh

"Rancho Zabaco Dancing Bull Zinfandel. First time I'd popped for non-jug wine in college and was blown away by the immediacy and hedonism that I had no idea was on offer in a bottle of wine."
Jim Bube

"The wine that sent me down a rabbit hole of obsession was a bottle of Domaine Roumier Chambolle-Musigny Les Amoureuses 1990 that my then-boyfriend, now husband, ordered early on in our courtship. We drank it in London in 2004 – pricing was different then!"
Jordan Salcito

"Pierre Gonon's Saint Joseph Vieilles Vignes. It was the first time I didn't have enough words to explain what I was experiencing in the glass. It was truly a transformative and emotional wine experience."
Justin Timsit

"Leth Roter Veltliner 1981 and 1986. Mature white wine that was screaming with vitality and freshness, with vibrant aromatics, ripping acidity, and textures unlike anything I had experienced."
Rachel Speckan

"I was bit by the wine bug on my first trip to Italy. I was told by a friend to find 1995 Brunello di Montalcino. I was hooked!"
Wendy Shoemaker

"My 'aha!' moment with wine was not drinking it but visiting Château Giscours in Margaux at the age of 13. I still remember the emotions I felt that day and the smell of the cellar."
Yannick Benjamin

Chartreuse

It's rare (in this book and in life) that
the category and the brand are one
and the same. But Chartreuse certainly
defies a genus.

"Herbal liqueur" is the unromantic category applied to it, but this
classification doesn't really do Chartreuse justice.

I was first introduced to Chartreuse, more as a lifestyle than
a mere spirit, at The Violet Hour. The Violet Hour opened in 2007
in Chicago's Wicker Park. The owners already had a few successful
restaurants in Chicago and partnered with New York-based cocktail
duo Toby Maloney and Jason Cott to create one of Chicago's first craft
cocktail bars. It became an instant phenomenon, every night attracting
a line of eager cocktail-seekers outside its heavy velour curtains.

Friends and I started to go, mainly on Monday and Tuesday
nights. These were the nights that nine-to-fivers would stay home
and us industry folks (as I now felt I was) came out to play. The
lines were less severe on these nights, and there was more time
to chat with the bartenders: always a handsome lineup of men
in dapper vintage vests (before this was a cliché), patterned ties,
and sleeve garters.

One night, emboldened by my third cocktail, I asked the bartender
about the omission. "So…why aren't there any girls behind the bar?"

Michael Rubel, as I would later learn his name was, looked at me with a mixture of fire and amusement. "Well, darling," he said in a way that felt neither misogynistic nor condescending, "none of them have been able to cut it." He paused. "I've been looking to get a woman behind this bar. I actually need someone part-time right now…if you know of anyone."

At this point Michael knew that I worked in the industry; but working at a wine store is a lot different than bartending. I didn't say anything, but affected a facial expression that obviously communicated *what about me?* This wasn't something I had previously thought about, but it suddenly seemed a foregone conclusion.

"You?" Michael asked.

"Yeah," I said, mustering the most confidence I could.

He examined my face for a good ten seconds, like the answer was in there somewhere. "Okay," he said. "Show up at 1pm tomorrow and we'll have a proper interview."

Our proper interview consisted mainly of him trying to scare me off the job ("it's hard, long hours on your feet, takes a lot of physical strength, there's a lot to learn," etc.) and me saying that I still wanted it. So he took a chance on me. And I started working two nights a week as a bartender at The Violet Hour.

Michael was right: it was not easy. It took a long time to build up the strength and endurance to be any good at it. There were hellish nights of punishing tickets at the service bar, of drunk guests hitting on me, of oppressive exhaustion. My sleep schedule was heavily disrupted, going to bed at 4am (at the earliest) two nights each week. My sleep issues became more severe: I sometimes woke up at 7am after going to bed at 4am; sometimes I could not fall asleep until 7am when I had to get up at 10am.

There was a time when I didn't think I would make it. I rehearsed what I would say to the partners in my resignation: that I was having some health problems, and that I just couldn't work these hours and at this intensity. Jason saw me struggling and offered me an out: one of the bartenders wanted to pick up more shifts. If I didn't want to be there, this was the time to say it. No hard feelings, no questions asked.

"No, I want to be here," I said, almost out of instinct. "I won't let you down."

This was the turning point. It would have been much easier to take the out. But I knew in that moment that I wasn't done at The Violet Hour. And once I had articulated this, everything became much easier.

Eventually, my nights at The Violet Hour became my favorite

of the week. I became good at working the bar and at formulating cocktails. I felt my palate developing in ways that had not happened yet in my wine studies. When I tasted a cocktail, I looked at its structural integrity: Was there enough acid? Too much? Enough sugar? Too much? Enough booze, or too much? Did there need to be more complexity, or did I need to simplify? Did the flavors make sense together? I started contributing recipes to the bar menu, and I curated the wine selections for the bar's list.

To be a bartender at The Violet Hour from 2008 to 2011 felt like being part of the 1927 Yankees (I googled "best sports teams of all time" to fill in this reference). Everyone was an all-star. Most nights around midnight, when we'd already been in service for six hours and had another two to go (not to mention at least another hour to clean up), we would take an "attitude adjustment": a shot that all the bartenders would do together to help us through the rest of the night. One bartender would line it up and choose the spirit — almost always an aged spirit or occasionally Chartreuse.

Green Chartreuse is 110 **proof** and made from over 130 herbs and botanicals. It has sweetness, but the burn of alcohol and bitterness wipes away any cloying notes. It is intensely celebratory or consoling, reserved for only the best and worst of nights.

The attitude adjustment was one of my favorite parts of the night, and it wasn't even about the booze. It was about acknowledging the work each of us had done and the sense of camaraderie that came with it.

—

The caliber of guests also made The Violet Hour an exciting place to work. I became good friends with cooks and sommeliers at Avec, Sepia, and Alinea. They trusted me to make them "whatever I was working on." We'd do shots at the end of the night, and when I went into their restaurants, they treated me like a VIP. As a 22-year-old just starting out in the industry, this was thrilling.

One night, a group of chefs from an important Chicago restaurant stopped by. Michael was entertaining these guys at the other end of the bar, and came over to me a little before midnight with a water glass in his hand. "Try this," he said.

As I was about to put my lips to the glass, I noticed there were large chunks of something floating in it. I looked at Michael, who was laughing hysterically.

"What the fuck is that?" I yelled. (I had started cursing more since bartending.)

"Oh, we were just doing a little shot of Green Chartreuse and someone's came back up," he said, still laughing.

To make matters worse, one of the chefs came to say goodbye when he left around 2am. He was drunk and didn't think much of kissing me right on the lips. I didn't think much of it either – until Michael told me he was the one who threw up.

Michael reached to the back bar and poured me a clean shot of Chartreuse. "Here," he said, "rinse it out."

46

Cocktails by The Violet Hour
Staff of 2008–2011

Whitney's With Me
by Andrew Mackey
The Violet Hour

Creme de Violette, to rinse glass
2 fl oz (60 ml) Chauffe Coeur VSOP Calvados
½ fl oz (15 ml) Dolin Genepy
¾ fl oz (22 ml) fresh lemon juice
½ fl oz (15 ml) simple syrup
3 dashes Boker's Bitters
orange peel, to garnish

Rinse a coupe with Creme de Violette. Shake the other ingredients with ice and strain into the glass. Garnish with the orange peel.

War of the Roses
by Mike Ryan
Kimpton Hotels and Restaurants

1½ fl oz (45 ml) Pimms No. 1
¾ fl oz (22 ml) London dry gin (Tanqueray preferred)
¾ fl oz (22 ml) St-Germain Elderflower Liqueur
¾ fl oz (22 ml) fresh lime
¼ fl oz (7 ml) simple syrup
2 dashes Peychaud's Bitters
2 sprigs mint, plus 1 leaf, to garnish

Thoroughly shake all liquid ingredients with ice. Add mint sprigs and agitate lightly, then strain into a coupe. Garnish with the mint leaf.

Tongue & Cheek
by Jane Lopes
Attica

1 strawberry, halved
2 fl oz (60 ml) Weller 107 Bourbon
¾ fl oz (22 ml) fresh lemon juice
¾ fl oz (22 ml) simple syrup
¾ fl oz (22 ml) Carpano Antica Sweet Vermouth
1 dash Angostura Bitters
2 sprigs mint, to garnish

Muddle strawberry in a shaker. Add the remaining ingredients and shake with ice. Place one mint sprig in a high-ball glass, cover with fresh ice and strain cocktail over. Garnish with the other mint sprig.

CHARTREUSE

Monkey's Heart
by Patrick Smith
The Violet Hour

48

1 orange slice
1½ fl oz (45 ml) Pisco
¼ fl oz (7 ml) Smith & Cross Rum
¾ fl oz (22 ml) fresh lime juice
½ fl oz (15 ml) Orgeat
1 barspoon St. Elizabeth Allspice Dram
1 dash Orange Bitters
mint and Angostura Bitters, to garnish

Lightly muddle the orange slice, then add the other ingredients, shake with ice, and strain into a rocks glass with fresh ice. Garnish with a bouquet of mint dressed with Angostura Bitters.

Art of Choke
by Kyle Davidson
Elske

3 mint sprigs, plus 1 extra, to garnish
1 fl oz (30 ml) white rum (preferably molasses based/not too grassy)
fat 1 fl oz (31 ml) Cynar
¼ fl oz (7 ml) Green Chartruese
1 barspoon fresh lime juice
1 barspoon Demerara Syrup (2:1 simple syrup made with demerara sugar)
2 dashes Angostura Bitters

Gently muddle mint sprigs in a pint glass, add the other ingredients, and let sit for 30 seconds. Fill with ice and stir, strain into rocks glass with a large ice cube, and garnish with a mint sprig.

The Pusher Man
by Eden Laurin
The Violet Hour

2 fl oz (60 ml) Lunazul Blanco
½ fl oz (15 ml) Green Chartreuse
¾ fl oz (22 ml) fresh lemon juice
1 fl oz (30 ml) pineapple juice
½ fl oz (15 ml) Demerara Syrup (2:1 simple syrup made with demerara sugar)
1 egg white
Soda water, to top up
Fees Old Fashioned Bitters, to garnish

Shake all ingredients without ice, then add ice and shake vigorously. Strain into a high-ball glass with fresh ice and top with soda water. Garnish with 5 drops of Fees Old Fashioned Bitters.

Tequila

Tequila is one of the few things I don't drink, through no fault of its own.

While working at The Violet Hour, I was asked by an important New York City bartender to be a brand ambassador for a Gran Centenario tequila called RosAngel. RosAngel is a Reposado tequila infused with hibiscus. It was designed to be a serious tequila, and Gran Centenario enlisted serious bartenders to represent it – though I'm sure a few marketing directors hoped that a pink tequila might attract a certain demographic.

They flew me out to New York for a weekend. The first day was a series of seminars on the production of RosAngel, as well as a show-and-tell for the cocktails we'd created for the event. One of mine was a play on the **Martinez** and the other was made with the proportions of the Art of Choke (see opposite page), but with ginger and Aperol instead of Green Chartreuse and Cynar.

We went back to our hotel to freshen up for the evening's events. Before leaving for dinner, the visiting bartenders convened in the hotel lobby for a brief drink. Someone decided it was a good idea to pour shots of tequila for the group. Although it occurred to me that it was going to be a very long night, I have a real problem saying no to shots; I love the sense of community.

Pre-dinner drinks turned into dinner turned into after-dinner drinks. I became friendly with a couple of New York-based bartenders (let's call them Robert and Brett) who wanted to show me some of the city's best cocktail bars. I was excited and flattered that they'd taken an interest in me. We went to Milk & Honey (in its original location), Death & Co., and Little Branch. When 4am rolled around and we wanted to keep going, I suggested they come back to my hotel room. I had a lot of tequila, courtesy of Gran Centenario, and "we could continue the party there."

They agreed, and we drunkenly piled into a cab. My hotel room was nice and modern but, as most things are in New York City, quite small. The only place to sit was the bed. The three of us jumped on and began drinking tequila straight from the bottle. They both started to casually stroke my legs. I was sober enough to be amused by the situation, but drunk enough to not really care where it led. Soon, Robert got up to go to the bathroom. Brett started kissing me. When Robert came back, we resumed drinking and chatting. Then Brett went to the bathroom and Robert and I started kissing. When Brett came back, I decided to go to the bathroom.

Holy fuck! I thought. *I'm going to have a threesome.* I had no idea what the logistics of this looked like.

I figured, when I came back from the bathroom, either one of them would be gone, or – if the night was heading in that direction – *they* would be kissing.

Sure enough, when I came back, Robert was putting on his shoes. So Brett and I spent the night together, or what was left of it. By the time we dozed off, the sun was rising. We slept for a good four hours before I had to check out of the hotel. Of the two of them, I was glad it was Brett who ended up staying. He was gentle and kind and I felt comfortable enough to tell him that he was the second person I'd ever slept with. He was flattered.

Since that night, tequila and I have not gotten along. It's like when you drink too much milk as a kid and develop a lactose intolerance. In one night, I'd gotten my lifetime supply. I still love the taste and the smell of tequila, but as soon as I drink any, I immediately feel unsettled.

I ended up developing many tequila drinks for the menu at The Violet Hour because I wanted to keep tequila in my life somehow. Ultimately, I'd had a lot of fun that night. And I felt like I was finally being a wild and stupid kid, something I'd never done throughout high school and college. Maybe I had been missing out.

50

Tequila by the Numbers

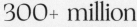

300+ million
the number of agave plants harvested in Mexico each year

274.1 million
the number of liters of tequila produced in Mexico in 2017

171.7 million
the number of liters of tequila exported to the US in 2017

1 million
the number of liters of tequila exported to Australia in 2017

1500
the desired altitude in meters for growing agave plants

199
the number of recognized agave species (tequila must be at least 50% blue agave)

65
the average weight, in kilograms, of an agave heart

9
the average age of a blue agave plant at harvest

1
the number of harvests that an agave plant yields in its lifespan

1
the number of animals that pollinate agave (bats!)

Somm Survey: Is there an alcoholic beverage that you categorically won't drink?

"Beer. Being a sommelier with Celiac disease definitely keeps me from engaging with the beer scene and the craft beer movement. The gluten-free versions are tolerable at best, and I usually gravitate towards cider if I am out at a casual bar."
Hannah Williams

"No, fully open mind – a reason for all seasons."
Caleb Ganzer

"Malt beverages. They are usually a candy-coated hangover."
Christy Fuhrman

"Four Loko. I'm scared."
Jonathan Ross

"I am loath to drink Malort as the bitterness is unbalanced, aggressive, and lingering."
Rachel Speckan

"I won't go near flavored vodka. Not only did I have a very bad experience with Raspberry Smirnoff at a Sublime cover-band concert when I was 19, I try to stay away from artificial or overly sweet flavors."
Rebecca Flynn

"Grappa! It's literally distilled from what is garbage or, at best, compost."
Richard Rza Betts

"Anything with Maraschino or Jack Daniels. There is something about the smell of those that my body rejects."
Theo Lieberman

"Anything with anise/licorice flavors. It's the one flavor I cannot stomach. Pastis, Sambuca, etc. Pas pour moi."
Victoria James

"I certainly appreciate Single Malt Scotch and I recognize its greatness, but the intense iodine and peat aromas I simply cannot handle."
Yannick Benjamin

"I have a real problem with cinnamon-flavored alcohol like Fireball. I believe it stems from a very bad experience I had with Aftershock as a child."
Desmond Echavarrie

"Vodka! It tastes like nothing!"
Wendy Shoemaker

Botrytis

I met Bo (not yet her name) at a pet shelter in Chicago.

She was a stray; her exact age is unknown. When I crouched next to her, she immediately jumped on my lap. She reached up and nuzzled my face, then made herself comfortable in the most temporary of spots: the plateau created by my squatted legs. She claimed me.

It had not been my intention to get a cat like Bo. I had wanted to get an older cat. Or a cat with a terminal disease. Or a deformed cat. I figured all the cute, young, affectionate cats could easily find homes. I would take a harder-to-love one. I looked at cats with Feline Immunodeficiency Virus. I saw cats who were too old to traverse stairs. I met a cat with one eye. But Bo meowed incessantly from the other side of the room, staring me down as I played with other cats. I vowed I would wait a month, and if she was still there when I went back, I would adopt her. If not, I would take it as a sign, and welcome a less lovable cat into my home.

—

A month later, I approached the shelter, nervous about not seeing my feline friend behind its doors. But she readily announced herself with a gentle and insistent chirp of a meow as soon as I entered the room. Bo did have a unique physical anomaly. Her left ear was clipped, creating a flat line where a triangle should be. This was done by the first shelter that found Bo – a spay and neuter clinic. They pick up strays, spay or neuter them, and put them back on the street. "We don't care what happens to you as long as you don't procreate" was the cold message. They clipped Bo's ear to signify that she'd been spayed: to ensure that she wouldn't be picked up again. I told myself that Bo's ear was enough of a deformation to count in my quest to love the unlovable. So she was the one I took home.

My friend Kelly was with me, and as we drove back to my apartment, I questioned what I should name my new roommate. Kelly, also in the wine industry, quickly volunteered Botrytis.

"Botrytis?"

"Yes, Botrytis. She's kind of sweet. Kind of rotten. Bo for short. It's perfect."

—

My parents had suggested I get a cat. Maybe they were responding to the tough times I was having – with men, with my health, with existential and professional angst. Maybe they knew that I had a hard time loving and being loved. For whatever reason, they offered the suggestion. And I bit.

When Bo and I got home from the shelter, I set up her food, water, and litter box in the bathroom of my one-bedroom apartment. (I was told you are supposed to introduce a new cat to your house one room at a time.) I sprawled on the bathroom floor and Bo jumped on my stomach. She laid her body across my torso and nestled her head in the crook of my neck. I distinctly remember feeling that I hadn't been so relaxed in a long time.

—

A few months after I got Bo, she became distant. Her voracious appetite subsided. She hid in the back of my closet. She winced at my touch. She lost a lot of weight very quickly. I took her to the vet, where they were noticeably alarmed. "I'm sure everything's going to be fine," they told me, but their tight smiles and somber tone said otherwise. They kept Bo while I went back to work. They called around 8pm to give me a status report. This time their words matched their intimations, as the doctor told me that Bo was stable for the moment and hooked up to IV fluids. He said

54

Bo had an issue with her kidneys, and they weren't sure she would pull through.

I was in the stock room of the wine store and couldn't catch my tears. This beautiful baby, who I'd only known for a few months, was sick. I couldn't do anything for her. And I might lose her. I thought of my own failing health and, in a moment of self-loathing, entertained the idea that my own energy had infected Bo, somehow making her sick. I cried for both of us.

I went home that night and drank Sauternes in Bo's honor. Maybe the rotten would defeat the sweet in my little Bo. I drank dessert wine until my teeth stung and my belly ached. I hoped that the Botrytis gods governing Bo would be appeased by my offering. I fell asleep on the couch without brushing my teeth.

I had a morbid dream about a cat drowning in a barrel of sweet wine. At first I saw the cat, lacquered and yelping, but could not help her. The next moment, I *was* the cat. I fell deeper and deeper, the light at the top of the barrel drifting further away. As the darkness almost consumed the light, I was blasted awake by a phone call. The sweetness of the wine and the saltiness of my tears still lingered on my tongue in an oddly harmonious way. Kettle corn for the soul.

"Ms. Lopes?"

"Yes…"

"This is Roscoe Village Animal Hospital." I braced myself for the worst. "Bo is doing very well. She has responded remarkably well to the fluids and is eating on her own. You can take her home tonight."

I thanked him profusely. And I silently thanked my Botrytis deities. The sweet had overcome the rotten.

Bo returned to the apartment that night. She was plumper, her eyes shiny, her step lively. She never hid in a closet again, and she never again looked the other way at a bowl of food (for better or for worse). She was back.

Bo slept in the crook of my armpit that night, her head resting on my chest. When I woke up the next morning, she climbed on my chest and rested her face on mine. She could only get closer if she put her head inside my mouth. In that moment, I knew that I could give love. And accept it.

BOTRYTIS

Botrytis (the Fungus) Explained by Botrytis (the Cat)

Bo can be sweet.

Botrytis, often called the noble rot, is responsible for creating some of the greatest sweet wines in the world.

Bo can be (selectively) rotten.

Botrytis Cinerea is the name of a fungus that afflicts grapes in certain growing regions. Sometimes it is highly desired, and sometimes it is heavily selected against.

Bo is lush and ample.

Botrytis dries out grapes on the vine, dehydrating them and concentrating the sugars, making them ideal for the creation of sweet wine. Botrytis also imparts a unique flavor. These wines end up being high in sugar, high in acidity, rich, viscous, and spiced with ginger, honey, and saffron.

VIGNETTE

Bo is charming and doesn't take no for an answer.

The most famous examples of botrytis are found in Sauternes, with Sémillon and Sauvignon Blanc grapes catching the mist of botrytis coming off the Garonne and Ciron rivers. Botrytis also shows up in the sweet wines of Tokaj, in northeastern Hungary, where the local Furmint and Hárslevelű grapes cannot resist its charms. The great Beerenauslese and Trockenbeerenauslese wines of Germany are made through the persistent spread of botrytis. It infects in Alsace and shows up in the Chenin Blancs of the Loire Valley. It even rears its head in regions it is deemed unwelcome: Burgundy, Sancerre, and Italy have all come to know botrytis from time to time.

Bo doesn't have to be sweet to get what she wants.

Botrytis-infected grapes don't always have to make sweet wine. Many of the great dry wines of Germany, Alsace, and Austria are made with a percentage of grapes that are infected by botrytis. The same flavor profile will apply, perhaps to a lesser extent, even in the absence of residual sugar.

The rotten and the sweet are at battle in Bo. Under the right conditions, and with the right amount of love, the sweet will prevail.

Botrytis is one of the great miracles of the wine world. Without the exact perfect proportion of morning humidity to afternoon warmth, it cannot exist. Too much of the latter and it does not grow. Too much of the former and it becomes an undesirable grey rot, destructive to the very vine that bears it.

Examination Wines

58

I'm not sure why I initially took the Introductory exam.

A few friends at work were doing it, so I jumped on board. I had no thoughts of ever becoming a Master Sommelier. It just seemed like the thing to do.

The Intro exam through the Court of Master Sommeliers is a two-day course that culminates in a multiple-choice test. The course is a whirlwind snapshot of the world of wine. Italy in 40 minutes. Spirits in 30 minutes. Eastern Europe in 15 minutes. The Masters hit the major points of interest in each region, more as a guide to future studying than an exact curriculum for the exam. Though, on occasion, a point of information is said with a wink, a gesture of *you should probably know this for the exam.* I remember the first growths of Bordeaux were said in such a manner. I made sure to learn them that night.

Though the Introductory exam doesn't test service or tasting skills, the material is introduced. There is a service demonstration. All I remember from mine was a Master carrying a tray loaded with way too many flutes and pouring Champagne excruciatingly slowly. Working retail at the time, I didn't have a clue how wine service worked, nor did I have any skills in that regard.

Blind-tasting was also a relatively new concept to me. In the exercise they set for us, each table "took" a wine that they would "blind" out loud to the whole group. One person on the table would stand up and describe the sight, the next would describe the nose, the third would describe the palate, the fourth would give an initial conclusion, and the fifth would provide the final conclusion. The sixth, seventh, and eighth people lucked out.

About three years into my wine career at this point, I could accurately describe wines…when I knew what they were. Having to describe a wine whose grape and region I did not know seemed like an entirely different skill to me, and one I had not yet developed. Then having to make a conclusion based on that description was even more daunting.

They got to my table and chose a girl three seats to my right to start. *Initial conclusion! Damn, the hardest*, I thought. After hearing the people before me describe the sight, nose, and palate, I would have to come up with a shortlist of what I thought the wine could be.

Luckily, I was a good test-taker, even if I wasn't a good blind-taster. I listened as the people before me threw darts at the wall to come up with descriptors. Then I watched the Masters' faces. When the descriptors were good, the Masters would look encouraging and hopeful, even nodding slightly with their eyes. *Yes! Go! You got it! You can do it!* When the descriptors were less good, the Masters looked down or to the side, an expression of benign disappointment on their faces. Instead of reading the wines, I read the Masters. Like a true psychopath.

"Red plum." That was a yes. "Cranberry, rhubarb, red cherry." All yeses.

"Blackberry pie." Nope. "Blueberry." Not really. "Craisin dust." That one wasn't a no or a yes, more a *why?*

"Vanilla." Yep. "Cedar." Sure. "Lilac." Why not?

"Graphite." Not really. "Saddle leather." I don't think so.

I collected that this was a red-fruited red wine. It was pale in color and I could see through it (I was sure about that one). It was aged in oak, thus the vanilla and cedar notes. It was a little floral. It wasn't terribly earthy.

"Umm…for my initial conclusion, I think this is a Pinot Noir from California," I said.

EXAMINATION WINES

The Masters had that hopeful and encouraging look, but also seemed to expect more. The woman next to me – the lucky girl who got to drive home my perfect pitch – nudged me. "You're supposed to say more than one thing."

But I couldn't think of anything else it could be. Nothing else made sense for the matrix of descriptors we'd assembled. I had to say something, though. "Umm…it could also be Pinot Noir from Oregon?" This was probably an unsatisfactory answer, but they let me move on.

I later learned that what I had just done was pretty much the deductive method. Only, it's supposed to be based on your own tasting observations and not the facial expressions of Master Sommeliers. Details. The wine turned out to be Russian River Valley Pinot Noir. And I felt pretty smug about that.

60

The Master Somm Process

Meet Mary. Mary is working in a restaurant to put herself through college when she first encounters fine wine.

She wants to learn more, but doesn't know where to turn, until she discovers the Court of Master Sommeliers.

She decides to sign up for the Introductory exam.

Level 1: Introductory Sommelier Course + Examination

+ Two-day course taught by Master Sommeliers to review the regions of the wine world, as well as spirits, beer, service and tasting
+ Exam is all written and multiple-choice
+ A score of 60% must be achieved

EXAMINATION WINES

Back at work, Mary asks to help out in the cellar. Her newfound responsibilities inspire her to pursue the next level of certification.

Level 2: Certified Sommelier Examination

+ Blind-tasting of two wines, a white and a red, and evaluating them accurately on a written grid
+ Written theory exam of multiple-choice, matching and short-answer questions
+ Practical service exam, which could include Champagne service or decanting, as well as **fortified** wines, spirits, cocktails and pairing
+ A score of 60% must be achieved

Mary's preparation for her exams helps her excel on the floor of the restaurant. She is promoted to full-fledged sommelier.

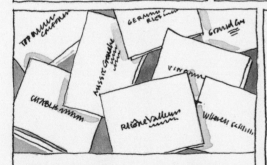

Inspired, Mary signs up for the Advanced exam. She ups her game, staging mock services with her friends, doing weekly blind-tastings and making thousands of flash cards.

Level 3: Advanced Sommelier Course + Examination

+ Three-day educational course is a prerequisite for the exam
+ Exam consists of three parts:
    1. Blind-tasting of six wines, three reds and three whites, in 25 minutes
    2. Written theory exam of short-answer questions and wine-list correction
    3. Practical service exam, which includes Champagne service and decanting, as well as beer, sake, fortified wines, spirits, cocktails, pairing and the business of wine
+ A score of 60% must be achieved in all three sections

GOOD TO SEE YOU AGAIN!

At the Advanced exam, Mary rekindles friendships with people she met at the Certified exam and forges new ones.

She passes and goes out to celebrate with her new friends.

Mary returns to work with a shiny green pin and is offered and accepts a position as wine director at a prestigious restaurant.

Mary's favorite part of her new job is mentoring new sommeliers, paying forward the knowledge given to her by others in the community.

NOW HOLD YOUR HAND OVER THE TOP.

CONS: TIME, MONEY, DIFFICULTY, STRESS.

PROS: KNOWLEDGE, EXPERIENCE, COMMUNITY, CHALLENGE, PRESTIGE.

But will she take the Masters?

Level 4: Master Sommelier Examination

+ Timed oral theory examination consisting of 80–100 questions; in the US, theory is given in advance and must be passed before candidates are able to sit for tasting and practical

+ Blind-tasting of six wines, three reds and three whites, in 25 minutes

+ Practical service exam may include Champagne service and decanting, as well as more advanced service techniques like double decanting, rinsing glasses, decanting from large format, service from small and large format bottles, beer service, and bartending; in the US, the practical section includes a business portion, and can also test the ability to blind-taste spirits, cordials, fortified wines and beer

+ A score of 75% must be achieved to pass each section

EXAMINATION WINES

Mary decides to take the Masters exam. She delves into the theory, studying for 20–30 hours per week.

She travels near and far to taste with Masters.

She stages mock services with other candidates, practicing the more difficult skills.

After a year of studying, she packs her bags for the exam.

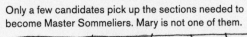

At the exam, she shares moments of fear and hope with some of the most talented sommeliers in the country.

Only a few candidates pick up the sections needed to become Master Sommeliers. Mary is not one of them.

Krug, as is turns out, does not taste like failure. It tastes like the promise of future success.

VIGNETTE

PRAGER *HINTER DER BERG GRÜNER VELTLINER FEDERSPIEL* Wachau, Austria
STROHMEIER *SCHILCHER FRIZZANTE* Steiermark, Austria
HEIDI SCHRÖCK *RUSTER AUSBRUCH* Burgenland, Austria

Austrian Wine

The summer of 2011, I found a delicate balance with my health.

My doctor put me on a custom-compounded porcine thyroid medication. Even though my labs didn't show traditional hypothyroidism, he thought it could help my symptoms. Slowly he increased my dosage. This, along with a less-common generic of Wellbutrin called Budeprion, brought me to a relatively stable place.

I was also at a turning point in my career. By now, I had been in Chicago for almost seven years. Three years in school and four working. I was ready for a change. It found me in the form of an opportunity to open a restaurant in Nashville, Tennessee. I jumped in, gave notice at both my jobs, and made plans to move south with Bo and my boyfriend of one year, Andy.

I spent the summer in Europe with my parents. We went on a tour of Austrian wine country, where we spent the day sifting gneiss between our hands and sipping Grüner Veltliner out of Zalto glasses. We took shots of pálinka before dinner in Budapest, and I made us whiskey smashes in our hotel room as we played cards overlooking the Vienna Opera House. I slept through the night. My stomach was relatively cooperative. I didn't have to think about my body or my health as much as I had for the past few years. I felt free. I had the space to worry about other things: *Was I prepared for my new job in Nashville? Was moving with Andy the right call?* I was thankful to be focused on my life instead of my body.

The Native Grapes of Austria

Austria has not traded in its indigenous viticultural heritage for internationally favored grapes. There are some non-native grapes – like Sauvignon Blanc, Pinot Blanc, and Riesling – that find unique and notable expression in the country. But this chart displays the grapes that can call Austria their motherland (or at least their adopted-at-birth home).

	BLAUER WILDBACHER	BLAUFRÄNKISCH	GRÜNER VELTLINER	NEUBURGER
Where it's grown	Blauer Wildbacher is almost entirely confined to Weststeiermark.	Distinctive styles are made in Carnuntum, Leithaberg, and Mittelburgenland.	Grüner Veltliner is the most planted grape in Austria. Half of Austria's acreage is in Weinviertel, and the rest in other regions of Niederösterreich and Wien (Vienna).	This natural cross of Roter Veltliner x Sylvaner is grown everywhere except Styria, but reaches its peak in the Wachau.
Styles it makes	The specialty of Blauer Wildbacher is a dry rosé called *Schilcher* (shill-chur). It can also be made in frizzante styles.	Peppery, slightly reductive grape varietal with good structure and length; the best ones taste like a cross between Syrah and Nebbiolo.	Known to make lean, peppery styles, as well as heavier, dry styles with elevated alcohol and **botrytis**. Because of its thick skins and loose bunches, the grape also does well for eiswein.	Robust dry styles with potential to age; floral, spicy, and nutty.
Wine to try	Strohmeier *Schilcher Frizzante*, Steiermark	Prieler *Blaufränkisch*, Leithaberg, Burgenland	Prager *Hinter der Burg Grüner Veltliner Federspiel*, Wachau	Weingut Lagler *1000 Eimerberg Neuburger*, Wachau

ROTER VELTLINER	ROTGIPFLER/ ZIERFANDLER	SANKT LAURENT	UHUDLER	WELSCHRIESLING	ZWEIGELT
"Roter" describes the red berries of this grape, grown mainly in the Wagram and Kremstal on loam or loess soils.	Two different grapes, but often said in the same breath as their blend is a specialty of Thermenregion, the village of Gumpoldskirchen in particular.	This native grape grows primarily in Thermenregion and northern Burgenland.	Uhudler is actually not just one grape; it refers to any "ungrafted" (i.e. native American) grapevine, common in Südburgenland.	Welschriesling is probably native to Italy but is considered an ancient grape in Austria. It is important for dry wines in Styria and sweet wines in Burgenland, particularly Neusiedlersee and Rust.	This cross of Blaufränkisch and Sankt Laurent, created by Fritz Zweigelt in 1922, is the most planted red grape in Austria. It can be found throughout the country, though it is most notable on the eastern side of Lake Neusiedl and in Carnuntum.
Dry styles, lean to rich with floral notes and more pink fruits.	Dry and sweet styles; nutty, spicy, and age-worthy.	Sankt Laurent is thought to be related to the "Burgunder" (Pinot) family and, as such, has a lean, elegant body. Different in its darker fruit quality and higher susceptibility to **Brettanomyces**.	Not recognized as the highest quality grapes, Uhudlers are nonetheless an important part of tourism and commerce in Burgenland. Sparkling styles are particularly successful.	Welschriesling has small berries with thin skins, making it the perfect candidate for botrytis. It is citrus-driven with good acidity, which compliments both sweet and dry styles.	Zweigelt ripens early, has good yields, and is not easily susceptible to disease. It is known for full-bodied, **barrique**-aged styles, but can also make fresh and lean versions, or even sweet wine.
Weingut Leth *Ried Fumberg Roter Veltliner*, Wagram	Fred Loimer *Gumpoldskirchner*, Thermenregion	Weingut Umathum *Sankt Laurent*, Burgenland	Uhudlerei Mirth *Uhudler Rot*, Burgenland (not very commonly seen outside Austria)	Heidi Schröck *Ruster Ausbruch*, Burgenland	Hannes Reeh *Zweigelt*, Neusiedlersee, Burgenland

FRATELLI BRANCA *FERNET-BRANCA* Milan, Italy
LUCANO 1894 *AMARO LUCANO* Bascilicata, Italy
MONTENEGRO *AMARO MONTENEGRO* Bologna, Italy

Amari

Nashville for me will forever be associated with Fernet–Branca.

It was a couple degrees of separation that got me the job. The Catbird Seat is owned by two great men and restaurateurs in Nashville: Benjamin and Max Goldberg of Strategic Hospitality. They also own The Patterson House, a craft cocktail bar on the first floor of the building that houses The Catbird Seat. Toby Maloney, one of the great barmen of our generation, helped open The Patterson House. Toby was also a part-owner of The Violet Hour.

Toby first told me about this project after a new cocktail tasting at The Violet Hour. Seasonally, when the menu at the bar would change, we would sit around and construct the new cocktails, talk about the intricacies of making them, and, of course, drink them. Those who were off that day enjoyed a nice 4pm buzz. Those who were bartending that night could look forward to a mid-service hangover. I was doing neither, instead rushing to my other job at the wine store LUSH. Toby stopped me on the way out, and in true Toby fashion, introduced the project by saying, "How would you feel about moving to Nashville?"

The more we talked about it, the more excited I got. I had been with LUSH for four years and at The Violet Hour for two and a half. There was no real possibility of advancement at either. Toby thought I would be a great candidate to run the beverage program at The

Catbird Seat because I had both an understanding of wine and experience with crafting cocktails. In his vision, both came into play in the beverage offerings at The Catbird Seat.

I met with Ben Goldberg when he visited Chicago. He is a kind and funny man who made me feel immediately comfortable. I met one of the head chefs, Josh, in Chicago over hot dogs. Josh was gentle and soft-spoken, while still being cunning and laser sharp. He was clearly passionate about the project and would fight for its success.

The Catbird Seat would be ambitious, for sure. The idea was a ten-course tasting menu that changed weekly, something Nashville had never seen before. The staff would be pared down, with only me, Josh, Erik (the other head chef), and two sous chefs cooking, serving, clearing, marking, and pairing everything. The experience would be intimate, with the person who cooked the dish serving it to the guest. Finally, and most importantly, it would be creative. Erik and Josh weren't about to create a restaurant that played by everyone else's rules. There would be room for play, to do things that hadn't been done before, and opportunities to create our own unique experience. I was in.

Ben's wife, Tara, helped me and Andy find a house to rent in Nashville, sight unseen. And on a sunny day in late August, we packed up a U-Haul, pumped Bo full of sedatives, and drove from Chicago to Nashville, Tennessee.

Moves involving copious amounts of wine have their own set of rules, so after the 11-hour drive, we had to spend another hour making sure none of the wine sat in the truck overnight. Josh, displaying our first-seen gesture of Southern hospitality, arrived at our doorstep at midnight to help unload a few things. That night we slept on a mattress on the floor. It was a darling house in West Nashville that would be ours for the next seven months, and mine for about seven more after that.

—

Neither Andy nor I had spent any time in the South. Nashville doesn't feel like the deep South, by any means, but it offers glimpses of that world, mixed with some old money, some new money, an important music scene, a burgeoning food and booze culture, several popular sports teams, a few big-name universities, and a booming medical industry. It all combined to make Nashville a unique slice of Americana, which I couldn't help but be charmed by. The people were friendly, their accents curling and rolling the same words that Chicagoans sharpened and cut. The city was beautiful. And the weather, especially compared to Chicago, was paradise.

Several days after the move, we had a team meeting. While I'd met with Josh and Ben several times now, the other three people I'd be working intimately with had yet to be uncloaked. We sat on the back porch of The Patterson House on a sweltering day and made our introductions.

Mayme and Tom would be the sous chefs. Mayme was Midwestern and had moved from Chicago to work at The Catbird Seat. She had previously worked at Alinea, which is how she knew Josh. Mayme specialized in pastry. She was young (most of us were), with undeniable talent and vision. Mayme would become one of my closest allies at the restaurant, and we shared many shots of rum over the tough moments in the opening year.

Tom was our local brigade, having grown up just outside of Nashville. He had cooked at The Patterson House, but besides that had no real culinary background. He was there to learn, and he proved himself every day in his determination and dedication. By the time I left the restaurant, he had become a talented cook in his own right.

And then there was Erik. He immediately struck me as someone I would let get to me. While Josh was soulful (for lack of a better word), Erik had a way of piercing my soul. No one at that restaurant would make me as happy and inspired as Erik did, and no one would make me as angry and frustrated. I had enjoyed peaceful and cordial business relationships with most people in my life, and Erik was the first person who made me want to throw things against the wall.

We decided it was a good idea to put these five people – Josh, Erik, Mayme, Tom, and me – inside a small, windowless room for 70-plus hours a week, throw in a whole lot of ambition, pressure, high expectations, and a packed reservation book, and see what happened.

It started out exciting. After all our hard work building the restaurant from the ground up, we were thrilled to see the vision realized. The doors opened in the first week to a parade of local food writers, musicians, football players, restaurateurs, chefs, and industry friends. Pretty soon, there was national press, and we had food writers from *GQ*, the *New York Times*, and *Food & Wine* in the restaurant.

My beverage offerings started receiving attention. Josh, Erik, and I envisioned that the pairings would be unique, offering up not just a standard glass of wine with every course. Beer, cocktails, teas, and juices would be incorporated. But I quickly started modifying wines, which hadn't to my knowledge been done before.

It began out of necessity: I didn't have access to the scope of wines that had been at my fingertips in Chicago. Sometimes I thought of a unique style of wine that would be great with a certain course, but it wasn't available in Nashville.

For example, Erik once created a slow-cooked pork dish with a sweet vinegar sauce. I thought the wines of Frank Cornelissen, a Belgian man in Sicily, would be perfect. He makes wines with little intervention, no sulfur, and wild and warm fermentations. The red wines are primarily made from Nerello Mascalese and taste like someone threw a bit of sour beer in red wine.

While Cornelissen wasn't available in Nashville, sour beer and Nerello Mascalese sure were. So I served a glass of Nerello Mascalese for the course. Red-fruited and ashy, with crushed velvet **tannins**, the wine was delicious on its own. Next to it, I served a carafe of Rodenbach, a Flanders red ale known for its sweet-and-sour, almost vinegary character, created by unique yeast and bacteria exposure. I explained my rationale to guests, how I came to this pairing, and invited them to pour as little or as much beer as they liked into the wine. Some people drank the wine on its own and drank the beer on the side. Others added a splash of beer, preferring the wine to dominate the flavor, with just a modest addition of tartness. And still others added the whole carafe, liking the wild and funky flavor that was after Cornelissen's heart. I wasn't so rigid to think that there was only one way to enjoy the pairing, and the element of interactivity in the meal seemed to engage people even more.

I started doing more and more pairings like this one. Soon, it became expected among our regular guests; if a particular pairing menu was light on such modified wines, they would jokingly comment on my lack of inspiration. One of the most beloved such pairings was a pour of **Tokaji** served in a wine glass rinsed with Black Maple Hill bourbon. This was paired with a pineapple, vanilla, and cherry dessert. I combined Alsatian Pinot Gris with **Falernum** to go with a passionfruit and foie gras dish. I made my own gin-based liqueur and topped it with dry, sparkling Muscat to go with another dessert. I combined Pedro Ximénez (sweet) and amontillado (dry) sherries to get the perfect level of sugar to match different soups. I poured sake in rosé Champagne

71

for fish crudo. A few guests and critics cried heresy – that wine was not meant to be manipulated like this. But most thought it was innovative and fun, and it became a signature of our pairings.

And the pairings became quite popular. Which made the act of service grueling. With sometimes up to 12 parties seated at the same time, most of them doing pairings, it was a struggle to maintain beverage responsibilities. It was a race against the clock to get pairings poured before the next course dropped.

The restaurant needed at least one more body. When we envisioned ourselves performing every aspect of service, just the five of us, we hadn't thought about most aspects of service. Greeting guests, discussing the menu format and beverage options, communicating any allergies or dietary needs, keeping water glasses full, clearing tables and resetting them for a second turn were tasks that fell hard on all of us.

The services were so taxing that by the time the last guests sat, and we got our feet under us, we needed something to help push us through the rest of the night. And so we resurrected The Violet Hour tradition of the "attitude adjustment." Josh, Erik, and I would hold court in the middle of the kitchen (thus, the middle of the dining room), pull an icy bottle of Fernet-Branca from the freezer and take a shot together.

Fernet-Branca is a nearly black, ruthlessly bitter amaro. Amari (as the Italian pluralize the word) are a class of Italian potable bitters, often drunk as digestifs. Some are sweet and bright, with only a delicate bitterness. Fernet occupies the opposite end of the spectrum, with only mild sweetness that is trounced by its bracing menthol character and bitterness. Sometimes we would toy with Amaro Lucano's consoling smoothness after service, or Amaro Montenegro's vivacious pep for a quick nip before, but the mid-service shot was always reserved for Fernet.

When we took that shot, the freezing cold Fernet felt like a minty ice-bath for my insides. Refreshing, shocking, and necessary. My mind and body became accustomed to the ritual, and the taste became associated with relief. *We had survived another night.*

Amaro Playlists

Music is a big part of what makes Nashville so special. You can walk into a random club or honky tonk on a Monday night and see some of the greatest music in the country. My music taste has always been somewhat pedestrian. I tend to listen to the same bands and songs over and over again: my brand of comfort food. Given this, I've enlisted some friends from Nashville to help create a playlist for three different amari. I recommend pouring yourself a glass neat at the end of a tough service and letting the songs play.

Fernet-Branca

1. **"This Must Be the Place (Naïve Melody)"** Talking Heads
2. **"The Funeral"** Band of Horses
3. **"Dirty Work"** Steely Dan
4. **"Bloodbuzz Ohio"** The National
5. **"The Beast and Dragon, Adored"** Spoon
6. **"Love Me Like a Reptile"** Motörhead
7. **"Goodbye Stranger"** Supertramp
8. **"Clapping Music"** Steve Reich
9. **"Black Star"** Radiohead

Amaro Montenegro

1. **"The Ghost Who Walks"** Karen Elson
2. **"Run Priest Run"** Wrekmeister Harmonies
3. **"Un Monumento"** Ennio Morricone
4. **"Rest"** Michael Kiwanuka
5. **"Sweet Child of Mine"** Guns N' Roses
6. **"Snowblind"** Black Sabbath
7. **"Who Am I?"** Les Misérables Original Broadway Cast
8. **"Vltava"** Bedřich Smetana
9. **"Big Mamou"** Waylon Jennings

Amaro Lucano

1. **"Thé à la Menthe"** La Caution
2. **"Into My Arms"** Nick Cave & The Bad Seeds
3. **"Under Your Spell"** Desire
4. **"¿Dónde Estabas Tú?"** Omara Portuondo
5. **"Black Velvet"** Alannah Myles
6. **"Let's Work"** Prince
7. **"Every Little Kiss"** Bruce Hornsby and the Range
8. **'Frank's 2000" TV'** Weird Al Yankovic
9. **"You Are"** Lou Rawls

1. **Max Goldberg** Strategic Hospitality, Nashville, TN 2. **Erik Anderson** Coi, San Francisco, CA
3. **Mayme Gretsch Walker** Hazel, Nashville, TN 4. **Robin Riddell Jones** Tavola, Nashville, TN
5. **Tyler Middleton** Graffam Middleton, Nashville, TN 6. **Colby Landis** The Catbird Seat, Nashville, TN
7. **Tom Bayless** Nashville, TN 8. **Patrick Halloran** Henrietta Red, Nashville, TN
9. **Tom Gannon** Spire Collection, Nashville, TN/New York, NY

Amaro Map of Italy

Three maps in one, this is first a geographic map of where the most famous Italian amari are made. Second, it maps the flavor characteristics of each. Third, their emotional profiles. Find what you are spatially, sensorially, and emotionally drawn to.

Ramazzotti Amaro, Canelli
+ Vanilla, liquorice, coffee
+ Stabilizing, warm, inviting

Amaro Nonino, Udine
+ Bergamot, tamarind, jasmine
+ Regal, complex, civilizing

Luxardo Amaro Abano, Padua
+ Clove, fruitcake, pistachio
+ Reliable, nostalgic, unifying

Nardini Amaro, Bassano del Grappa
+ Dark chocolate, peppermint, bitter orange
+ Commanding, gentle, smooth

Amaro Montenegro, Bologna
+ Cucumber, rose water, clementine
+ Delicate, light-hearted, lively

Fernet-Branca, Milan
+ Menthol, bay leaf, tea tree oil
+ Pungent, cleansing, bracing

Amaro Bràulio, Valtellina
+ Pine resin, chamomile, fig
+ Invigorating, lean, alpine

Rabarbaro Zucca, Milan
+ Rhubarb, black cardamom, charred red peppers
+ Rejuvenating, rambunctious, raw

Amaro di S. Maria al Monte, Genoa
+ Ginseng, radicchio, spearmint
+ Powerful, brisk, taming

Amaro Sibilla, Macerata
+ Honey, smoke, dandelion
+ Shocking, dynamic, purposeful

Amaro Meletti, Ascoli-Piceno
+ Saffron, anise, milk chocolate
+ Affirming, subtle, pure

Amaro CioCiaro, Sora
+ Orange oil, cola, cinnamon
+ Peppy, simple, encouraging

Amaro Lucano, Pisticci
+ Fennel, black tea, burnt orange
+ Balanced, sunbaked, consoling

Averna Amaro Siciliano, Caltanissetta
+ Mocha, citrus peel, star anise
+ Brooding, sultry, soothing

Piedmont

Lombardia

Veneto

Friuli

Liguria

Emilia-Romagna

Marche

Lazio

Bascilicata

Sicily

PETRUS *GRAND VIN* Pomerol, Bordeaux, France
MASSETO Tuscany, Italy
VÉRITÉ *LA MUSE* Sonoma County, California, USA

Merlot

There was a romance and a glamour to Nashville, even through the hard services.

But there was a dark side too. The calm that I had enjoyed the previous summer was disturbed in Nashville. All the problems, the sense of unwellness, came back with a vengeance.

I remember being in Whole Foods when I first felt it creeping back into my life. (I'm sure there's some sort of parable here about how what we do to fix our health often makes it worse.) It was less of a creep, though, and more of a wave. I couldn't see straight. I felt dizzy. I held onto the shopping cart to keep me upright. All those familiar feelings of fear and loss of control rushed over me, and I immediately clenched myself in tight. The freedom and peace that had surrounded me over the summer – gone, in an instant.

I didn't change anything immediately, but I hoped that if I remained constant, everything would go back to the way it had been. I didn't have any doctors in Nashville, and I didn't have time to go chasing my ailments. I was the only one at The Catbird Seat who could do my job, and it was important that I did it with expertise and style. I couldn't be seen as weak or faltering.

I was the beverage director, but also the only real front-of-house employee. In the beginning, I folded the napkins, polished the silverware and glassware, organized the floor plan, handwrote the tickets for the kitchen, filled water bottles, wiped down the seats and swept before service, and set the entire room, in addition to the responsibilities of running the beverage program. I could have dealt with the exhaustion, but I had a hard time dealing with the self-doubt. I felt like I had no idea what I was doing, and that one day someone would find out what a fraud I was.

Sleep, again, was the first thing to go. Even when I did sleep, I would wake up in a state of panic that pervaded my entire day. Initially, my work days ran from 11am to 3am. I would wake up early to do yoga, just to help me feel a little more grounded, working on just a few hours of sleep most days. My stomach was a wreck and I drank way too much.

I didn't want to find a new doctor in Nashville so I called my old psychiatrist from Chicago. Budeprion was no longer being produced, so he called in a prescription for Klonopin, which I had previously responded somewhat well to.

Klonopin is a drug in the benzodiazepine family, historically prescribed for seizures. It is a sedative that has strong indications for tolerance and withdrawal. One-third of people who take Klonopin for more than four weeks grow dependent. Withdrawal can take years, and symptoms related to withdrawal can persist for years after that. I knew none of this.

When I first started taking it, my mornings improved. I did not wake up in my usual state of panic, for which I was thankful. But I would have more severe sporadic panic attacks, to which my doctor responded that I needed to take more Klonopin. I remember being in a hotel room with Andy on a visit to Chicago, crying uncontrollably and having no idea why. *Just take more Klonopin* was the email I received from my doctor. And I did.

I started crying all the time. Many days, I found myself in the bathroom of The Catbird Seat before service, wiping my eyes and pulling myself together. I blamed my circumstances: I was too stressed, I worked too much, the chefs didn't get me menu descriptions in time, my distributors always messed up my orders, I worked in a windowless room for 70 hours a week. It never occurred to me that it was the Klonopin making me cry. I thought it was the only thing that was holding me together.

Eventually my Chicago psychiatrist stopped prescribing for me. He said I needed to find a local doctor. I vaguely remember my new doctor telling me about the dangers of long-term Klonopin usage. Per usual, I didn't fear that the future could be worse than the present. The Klonopin was providing me relief, and I continued to take it.

At this time, I began seriously studying again. Hard times always pushed me back to the books. I figured, if I couldn't be relaxed and happy and feel good, I could at least be productive. I made flash cards and study guides in preparation for the Certified exam through the Court of Master Sommeliers.

I spent the night before the exam cooped up in a Cincinnati hotel, crying my eyes out. I gasped for air and shook uncontrollably as I eventually wore myself out and fell into a few hours of sleep.

—

The theory portion of the exam was well within my realm of knowledge. I remember not knowing the answer to only one question: In what appellation is Château Clinet a producer? I guessed Saint-Émilion and later looked up that it was actually Pomerol, a prestigious appellation for Merlot-based wines on the right bank of the Dordogne River in Bordeaux. Coincidentally, I also called my blind red wine that day as being California Merlot. I remember it being soft and fruity, and that's about it. It could have been Beaujolais, it could have been Zinfandel. It was most likely not California Merlot.

My service portion was relatively smooth, though at this point in my career, my tray-carrying skills were limited at best. At LUSH and at The Violet Hour, I never had to carry a tray. And at The Catbird Seat, without any guidance, mentorship, or supervision, I didn't force myself to do the hard things. If a four-top was starting with four glasses of Champagne, I would pour them at a side station and carry two stems between my fingers in each hand. If a six-top ordered six glasses of Champagne, I would make two trips.

I ended up passing that day, with the second-highest score. The Masters were very complimentary of my theory. My service, they said, was strong, but I "looked like I was going to drop that tray." And I got a few notes on areas to improve my tasting, including "red varietal identification."

I left that day knowing two things:

#1   I had to move out of Nashville. If I wanted to be serious about wine, I needed a serious community of people who could push me, hold me accountable, and teach me how to use a tray. I needed to work in a restaurant where I wasn't the person who knew the most about wine.

#2   I did not understand Merlot.

MERLOT

The World of Good Merlot

KNOWN FOR

Washington State, USA
Windy conditions creating structured, deep styles

California, USA
Plummy and soft styles, when compared with the ubiquitous Cabernet Sauvignon

Long Island, New York, USA
Bright, herbaceous, and clean styles that don't rely heavily on oak or extraction

Bordeaux, France
The definition of luxe, concentrated, earthy, and oaky styles

Switzerland
Fruity and clean styles, with some oak presence, as well as WHITE Merlot (made without skin contact)

Friuli, Italy
Stern styles with pleasant bitterness, purple fruit, and lots of herbaceousness

Tuscany, Italy
Blockbuster styles with plenty of ripeness and oak, yet enough **VA** and structure to remind you that you're in Italy

Bulgaria
Clean and herbal styles, lighter in body, with tart fruit

Hawkes Bay, NZ
Serious and age-worthy styles that are red-fruited, structured, and complex

South Australia, Australia
Plum, mulberry, and sage from classic producers and lighter, more experimental styles from the new wave

Chile
Merlots that have been confused with Carménère, and still lack premium status, yet are forging an identity of their own

START MODEST

L'Ecole No. 41 *Merlot*, Columbia Valley

Duckhorn *Merlot*, Napa Valley (modest by
Napa standards)

Macari *Reserve Merlot*, North Fork of Long Island

Château Grand Village, Bordeaux Supérieur

Paolo Basso *Il Bianco di Chiara*, Ticino
(white Merlot!)

Venica & Venica *Merlot*, Collio

Poliziano *In Violas Merlot*, Cortona

Rossidi *Nikolaevo Merlot*, Thracian Valley

Villa Maria *Private Bin Organic Merlot*

Brash Higgins *MRLO*, McLaren Vale

Errazuriz *Estate Series Merlot*, Aconcagua Valley

OR GO BIG

Leonetti Cellar *Merlot*, Walla Walla Valley

Vérité *La Muse*, Sonoma County

Wölffer Estate *Christian's Cuvée Merlot*,
The Hamptons – Long Island

Petrus *Grand Vin*, Pomerol

Gialdi *Sassi Grossi Merlot*, Ticino

Miani *Filip Merlot*, Colli Orientali del Friuli

Masseto (formerly produced by Tenuta dell'Ornellaia,
now its own entity)

Nothing's too big in Bulgaria yet!

Craggy Range *Sophia Merlot*, Gimblett Gravels

Irvine *Grand Merlot*, Eden Valley

Casa Lapostolle *Cuvée Alexandre Merlot*, Apalta,
Colchagua Valley

MERLOT

Barolo

I can't remember the first time I had
Cordero di Montezemolo.

I was given a bottle by a distributor when I was working in Chicago.
I do remember it was from the 2004 vintage, but I don't remember
when I drank it – a testament to my lack of context in the early days
of my wine career.

Barolo is a small region in Piedmont, in northern Italy. More Alpine
than Mediterranean, more butter than olive oil, more backwoods than
cosmopolitan, it is the source of some of the finest red wines in the
world. The grape is Nebbiolo. It is thin-skinned, producing wines
of transparent ruby-garnet color, belying their power and intensity.

When I drank Cordero for the second time, it was chosen for
me by a sommelier at Eleven Madison Park named Jonathan Ross –
Jon. It was the first time I had eaten at the restaurant, fall of 2013.
I had just returned from a wine trip around Germany and Italy, had
passed my Advanced exam through the Court of Master Sommeliers
the month before, and was approaching my 28th birthday. To
celebrate, I went with my best friend from childhood, Megan, and
her mother, Debby, to Eleven Madison Park. Jon and I had been
seeing each other (casually, off and on – it was complicated) for
the previous six months.

Jon wheeled the bottle – vintage 1989 and a magnum – out on
a **gueridon** that carried an unexpected line-up of props: a butane

burner, a shaving brush, a coffee siphon clamp, and a pair of tongs. He proceeded to behead the bottle – cracking the glass just below the cork. It was an impressive show, and a more impressive wine.

Great Barolo is like a tuning fork. It picks up on the vibrations and energy of those drinking it and amplifies them. No wine is so emotional or dramatic. The stakes are never higher than when you are drinking Barolo. Everything feels like a miracle. The three of us talked and laughed, plotted for the future and reminisced about the past, falling deeper into the spell of the wine.

Later that night, while we were lying in bed, Jon told me that his co-workers all commented on how beautiful his girlfriend was. I wasn't his *girlfriend*. Despite myself, though, I wished that I was. I wasn't sure why he made that comment, but it gave me hope. I would sew together small threads like this in the fantasy that one day they might keep me warm.

He turned over on his stomach and fell asleep. I lay on my side, facing him. A whisper-thin line between us felt like a magnetic field, with my positive unable to attract his negative. I held my pillow and finally dozed off, my emotional excitability losing out to the haze of Barolo and a good meal.

—

I met Jon before I moved to New York. My friend Tom brought me to a tasting group at Del Frisco's Steakhouse while I was up visiting for a weekend in January of 2013. I was still living in Nashville at the time. I gave my notice at The Catbird Seat after about a year of working there. I had broken up with Andy months before, lumping him in with what I believed to be the circumstantial causes of my depression. And for a time, I did reach a lull in the storm. My body had become accustomed to the Klonopin, and I sailed through the next few months with little emotional turbulence. I moved to New York City with the plan to find a job in wine and study for my Advanced exam through the Court of Master Sommeliers.

Though I don't recall especially noticing Jon at that tasting, I remember the wine he blind-tasted (Haut Brion Blanc) and what he called it (white Rioja), so he must have made an impression. I did have a new boyfriend at the time who lived in Nashville, a good friend who became more when my bags were already packed. He and I exchanged emails of epic length, which he joked would one day become the next Great American Novel. I was enchanted by the romance of it all, but there was always an itch that we wouldn't ultimately be compatible.

When I moved to New York, I continued to see Jon in tasting groups. There was something about the way he moved. He was cocky and assertive, yet he moved with grace and precision. I caught myself watching him walk around a table with a tray full of glasses, lingering a little too long before averting my eyes. Jon was clearly passionate about wine, funny, irreverent, and he was always the first to offer his time and energy to others.

A seed was planted, but I had not yet registered it as a crush. I did know, however, that I did not want my current boyfriend to move to New York. I proceeded to initiate an emotional breakup with a man who had already made plans to uproot his life for me. I cried through the whole thing but felt immediate relief when it was over.

It was April in New York. I was single and felt a lightness I hadn't experienced since moving to the city. Of all the bars I could have sat at that night, I decided to do something nice for myself, and go sit at the bar at Eleven Madison Park. And what a glorious bar. A study in contradictions, it managed to be both comfortable and elevated, intimate and grand, raucous and peaceful. I sat down, ordered a drink, and exhaled – for what felt like the first time in months.

I would look back on the moment that followed with some regularity. Jon emerged from around a corner and saw me. His cheek creased as a smile emerged. His eyes shone. He titled his head inquisitively, silently asking "what brought you here tonight?"

And at that moment, I knew what had brought me there.

Modern vs. Traditional Barolo (and Barbaresco)

Barolo and Barbaresco producers are often talked about as being "traditional" or "modern." Most sommeliers recognize that there is a spectrum of modernity, and I would go even further to say that there is a spectrum across multiple facets of production that interact to create a unique profile for each winery. The chart on the following spread maps two of the most significant factors in placing producers within this spectrum: oak usage and maceration length.

Traditionally in Piedmont, grapes endure a lengthy and gentle maceration on their skins to extract **tannin** and flavor (with the added effect of reducing color). Modern thinking shortens this maceration time to maintain color and soften tannins, but often uses more aggressive techniques to extract as much flavor as possible in a short period of time.

Traditional wines age in old **Slavonian botti** while a modern approach incorporates new **barriques**. The gradients within the oak spectrum are detailed on the chart.

There are further nuances to the modernity conversation. Some producers have evolved their style over the decades; the chart reflects current winemaking. Some producers treat specific wines with a more or less modern touch; the chart reflects the major house style. Several other factors can contribute to a modern or traditional style other than these two: length of time spent in oak, fermentation vessel, **cap management**, temperature control, and **clonal selection**.

I encourage building up a repertoire of the producers you like and seeing what choices they make in common. I also encourage performing a blind-tasting with several modern-leaning and traditional-leaning producers from the same vintage. I did this and was blown away by how much pleasure I took from both ends of the spectrum.

Modern vs. Traditional Barolo (and Barbaresco)

OAK USAGE

3 Days

SKIN MACERATION LENGTH

OAK USAGE	
Made with the purchase of new oak each year, beginning with a small percentage and rising to a very large portion of new oak: by design, those using new oak would not be purchasing barrels larger than 500 liters each year (this also causes the need to use alternative fermentation vessels like stainless steel and concrete)	Elio Altare La Spinetta Scavino Domenico Clerico Ceretto R. Voerzi Sandron
Mix of medium and small barrels (500-liter tonneaux and smaller) but mainly from used wood of various sources	Cordero di Montezemolo
Mainly medium-sized oak barrels (less than 5000 liters but greater than 500 liters), from used wood of various sources	Renato Ratti
Mix of medium and large casks, old, and including a mix of oak sources	
Old, large botti made of French oak, or any oak other than Slavonian; all fermentations occur in oak or concrete	A. Conter
Old, large botti made from Slavonian oak; all fermentations occur in oak or concrete	

VIGNETTE

The most traditional wines of Barolo (bottom right) tend to be tannic, pale in color, and rustic. The most modern wines (top left) tend to be dense and darkly colored, with more pronounced purple fruit, oak flavors, and softer tannins. But there is a spectrum among multiple factors that determine a wine's flavor profile; use this chart as a jumping-off point to discover which Barolos you are drawn to.

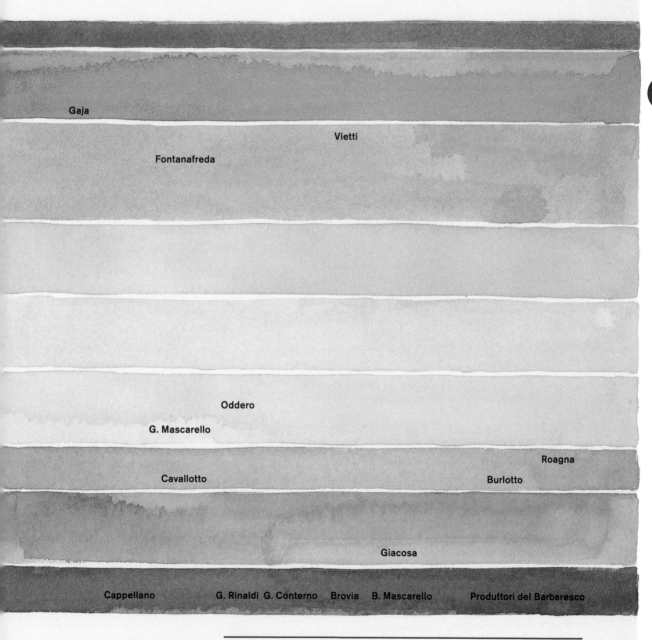

Gaja

Vietti

Fontanafreda

Oddero

G. Mascarello

Roagna

Cavallotto

Burlotto

Giacosa

Cappellano   G. Rinaldi G. Conterno   Brovia   B. Mascarello   Produttori del Barbaresco

BAROLO

Vinos de Jerez

Following our encounter at Eleven Madison Park, Jon and I spent the night together.

Our first of many. When we were together, my body hummed. My eyes were clear. My smile was brighter and unrestrained. We sat on the floor and drank palo cortado sherry. We talked about wine and work. But when I asked him about his family, his past, his future, he was hesitant. He receded from me, answering quickly and not reciprocating my curiosity about him. I wrote it off. I made excuses. I figured I was just moving too fast and pushing too hard. I wanted to spend time with him, no matter what doubts and insecurities bubbled under the surface.

Palo cortado is a type of sherry (or vino de Jerez, to un-Anglicize it) that was originally a product of a spontaneous transformation. Dry sherry can largely be split into two categories: the first is aged under a layer of yeast known as flor; the second is aged oxidatively, without any yeast for protection. Palo cortado would begin its life under a layer of flor, which would (historically) disappear spontaneously. Then the style would age oxidatively. Its characteristics would include the green-almond and saline freshness of an unoxidized sherry with the caramelized notes and full body of an oxidized one. But its ultimate cachet was this sense that it was sort of a miracle to even exist. *Why did the flor disappear on this wine? What is so special about this one barrel?*

—

Jon and I saw each other every week or so, each time with an awkward lead-up – me posturing to not seem overly excited, and him guarded and reticent.

Somewhere in these initial few weeks of our relationship, Jon took the Master Sommelier exam for the first time. He had gotten off the wait-list six weeks before and was wholly unprepared. I gave him mock tests, which he bombed, and we joked that it would be a good trial run. This was 2013 and the last year that all three sections of the American MS exam were offered together. Jon thought he had a good chance at the practical portion. (I think his exact words were: "I'll quit my job if I don't pass service.") After all, he was a sommelier at Eleven Madison Park, and worked in a technically demanding service environment every day. We texted and talked the whole time he was in Aspen for the exam, and I comforted him when he found out he didn't pass any of the sections.

We met up the night he got back. His flight was delayed and he took a cab straight from Newark Airport to meet me at a bar in the West Village. He walked in the door, put his suitcase in the corner, grabbed my face, and kissed me. My whole body heaved with energy. I felt like everything would be better now. It was just the exam that had put him – and our relationship – on edge. We went home together that night. I slept poorly, waking up periodically to his chest rising and falling, and tried to insert myself between it and his arm. He remained in deep sleep, untouched by my restlessness and unmoved by my warmth. His arm darted out straight, never curling in to embrace me. And I knew that everything was not better now.

I tried – briefly and unsuccessfully – to keep things breezy and unemotional with Jon. But ultimately, I was not content to be casual. Everything he gave me made me want more. I had told three men before Jon that I loved them, but I had never felt this way before. My patience had always been a thin veneer that held no contest against my ballooning passion and persistence and, in this particular case, it barely put up a fight.

One night Jon and I met up late at Milk & Honey (in its second location on 23rd Street): one of those bars that feels like it belongs in a different era, with moody lights and stiff drinks. I ordered a **McKittrick Old-Fashioned** and he ordered a **Bamboo**. I told him I was frustrated with him. That I felt he was unengaged and didn't make an effort for me. And I told him that I was falling in love with him. He went home with me that night but didn't stay. He told me he could see himself one day feeling the same way, or else he "wouldn't be here." We made plans to see each other the next night, after he got off work.

The following night, 1am came, and not a word. I knew he was working, but I thought I would have heard from him by now. Perhaps a few comforting words after my confession the night before? Perhaps an affirmation that he was excited to see me? Nothing. I finally sent an angry text, something – now lost to history and technology – to the effect of *Why do you make promises that you don't keep?* He shot back after a few minutes, explaining that he'd just gotten off work and *Why are you being so crazy?* A few more exchanges back and forth, which ended with him rescinding what he'd said the night before. *I could never fall in love with you.* I threw my phone against the wall and collapsed into tears.

—

These days, palo cortado is not created through a spontaneous metamorphosis. When a cellar-master in Jerez sees a barrel of fino (the unoxidized type of sherry) that is particularly robust, he or she might **fortify** it. At a heightened alcohol level, the flor that was protecting the fino from oxidation dies off, and that wine now ages oxidatively. It will become a palo cortado. The wines can be some of the most spectacular in Jerez, but they are a decision, not a miracle.

The Myths of Sherry

Myth: It's sweet.

Truth: Most sherry is made in a *vino generoso* style, which means it has less than 5 grams per liter of **residual sugar**. Because the activity of the flor consumes glycerin, these wines taste even drier than normal table wine, and are quite refreshing, even given the elevated alcohol.

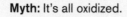

Myth: It's all oxidized.

Truth: Even wine professionals often describe biologically aged sherry (i.e. fino and manzanilla styles) as having "nutty" and "oxidized" notes. Biologically aged sherry ages under a layer of yeast called flor, which completely protects it from oxidation. The unique green-almond, saline, and curried character that develops in these wines is not a product of oxidation, but rather of sotolon, a flavor compound developed through the yeast contact.

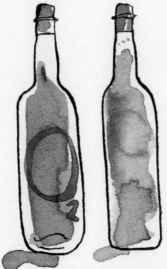

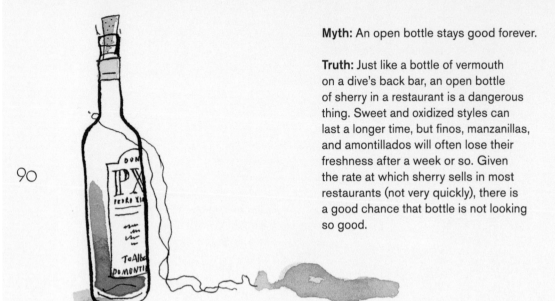

Myth: An open bottle stays good forever.

Truth: Just like a bottle of vermouth on a dive's back bar, an open bottle of sherry in a restaurant is a dangerous thing. Sweet and oxidized styles can last a longer time, but finos, manzanillas, and amontillados will often lose their freshness after a week or so. Given the rate at which sherry sells in most restaurants (not very quickly), there is a good chance that bottle is not looking so good.

Myth: A solera looks like a pyramid of barrels.

Truth: The solera system of aging is designed to fractionally blend older wine with younger wine. In Jerez, it serves the dual purpose of distributing the complexity of old wine across a large volume, as well as maintaining the vitality of the flor (yeast) activity through the addition of young wines. The different tiers of a solera are often spread across different rooms. If they are stacked at all, it is often a combination of different soleras, with finos liking the cooler and more constant temperature close to the ground, and olorosos tolerating the heat at higher levels.

PEGASUS BAY *RIESLING* Waipara Valley, Canterbury, New Zealand
BELL HILL *PINOT NOIR* Canterbury, New Zealand

Kiwi Wine

It took some time to feel at home
in New York.

In the first months of living in the city, I stayed with my friend
Megan's mom, Debby. I was planning to move into my friend Tom's
spare bedroom in Harlem, but Debby insisted I live with her. She
made Bo and me feel at home in her beautiful Upper East Side
apartment, and never asked for a thing in return. I stayed there until
the lease on Megan's studio was up, and we found a two-bedroom
apartment in SoHo together.

New York infantilizes like none other. I hadn't lived with a
roommate – let alone taken charity lodging – in years. I had made
a good living in Nashville and Chicago and had afforded myself
beautiful apartments in which I lived alone. In New York, rents
are high and competition fierce. The jobs I was qualified to do in
previous cities were out of my reach here. The apartments that were
previously in my budget were reserved for the truly wealthy. Living
alone in New York is a luxury.

Megan and I were happy with our space. It had high ceilings,
modern appliances, and was just on the eastern cusp of SoHo.
It had no oven (the landlord thought it attracted vermin), but Megan
and I rarely cooked. You could've bought a palace in Chicago or
Nashville with what we paid in rent, but that's the sacrifice you make
to be in New York.

I moved my things from the Upper East Side to SoHo in a taxi, Bo squealing in a cage far too small for her ample size. It was the day after Jon told me that he could never fall in love with me. I texted my sister about our conversation. Beth called him a dick and told me to forget about him. Most people who knew me figured that I didn't actually love Jon. How could I love a man who had such ambivalence for me? They figured he was only attractive because I couldn't have him.

I cried the entire car ride and left a very generous tip for the driver. I lay down on the floor of my new apartment and waited for my other things to be delivered from storage. *This is the moment*, I thought, *that screw-caps were invented for.*

I had purchased a couple of bottles of New Zealand wine to aid with my studies. I remembered reading that New Zealand was such a geologically young country that the sites with the oldest (i.e. poorest and best) soils were on flatlands and valley floors. The hillsides, which tend to be the premium sites in most other countries – countries old enough to have experienced hundreds of thousands of years of erosion – are the young and fertile sites in New Zealand. A true anomaly in the wine world.

I sat cross-legged on the kitchen floor and pulled the wines out of my bag. I unscrewed each: a Riesling and a Pinot Noir from Canterbury. Call me pretentious, but in general I do enjoy drinking wine out of a glass. Something about aromatics, openness, *blah blah blah*. But I made an exception on this occasion, taking a few giant swigs out of each bottle.

I was drinking in part because I was sad about Jon. Moreover, I realized, I was drinking to christen our new apartment. I was grateful to be in New York, to have a home of my own, and I was even grateful for Jon. Even if he never loved me, *how cool is it to feel this strongly for someone?* The wines were a perfect accompaniment to the moment: fresh and unassuming, young and ambitious, comforting and challenging. They seemed to say *even though things may be different down here, we're just as good (if not better) for it.* They made me feel I was where I should be.

An Illustrated Guide to Closures

Cork

Pros: elastic, natural, romantic, proven ageability
Cons: cork-taint and failure rate with low-quality styles, expensive

Diam

Pros: minimal risk of cork-taint, maximum romance
Cons: some note a "gluey" flavor, still some risk of cork-taint and oxidation, ageability yet to be proven

Plastic/Synthetic

Pros: inexpensive, fun colors, no cork-taint
Cons: not good protection against oxidation, poor choice for anything intended to age more than a year or two

Screwcap/Stelvin

Pros: inexpensive, high-quality seal, no cork-taint, easily resealable
Cons: accused of being too airtight and not providing proper development in the long term, zero romance

Glass/Vino-Lok

Pros: looks elegant, easily resealable, no cork-taint, is meant to mimic the oxygen ingress of cork to allow proper aging
Cons: relatively expensive, not a lot of examples of proven ageability, still lacks the romance of cork

Crown Cap

Pros: looks cool, inexpensive, holds pressure well (often used for sparkling styles), no cork-taint
Cons: no proven ageability, a different sort of romance

The Geographic Features of NZ

The Māori are the *tangata whenua* – indigenous people – of New Zealand. They inhabited "the land of the long white cloud" for centuries before European settlers arrived. They still make up a significant percentage of the nation's population, and their language, traditions, and history heavily influence New Zealand's culture today.

The Māori have a special relationship with the geographic features of New Zealand. Their folklore and oral storytelling centers on the land, soil, and water of the country. These same features are what create such a unique viticultural landscape.

Anyone who has studied New Zealand knows the pain of distinguishing between Waipara, Wairarapa, Waikato, and Waitaki. The linguistic foundations (below) of many of these placenames give context and help differentiate these often similar-sounding names.

The facing map shows some of the country's major geographic features. Regions are outlined, with the key wine-growing areas shaded within each.

+ **Au:** current
+ **Ara:** path
+ **Awa:** river
+ **Heke:** to ebb, drip or descend
+ **Iti:** small, little
+ **Kato:** to flow
+ **Kume:** to pull
+ **Koko:** corner, or to shovel
+ **Mānia:** plain
+ **Manga:** stream
+ **Mata:** face

+ **Maunga:** mountain
+ **Moana:** sea, or large inland lake
+ **Motu:** island
+ **Ngarue:** to shake, as in an earthquake
+ **Nui:** large, big
+ **One:** sand, earth
+ **Pae:** ridge, range
+ **Para:** mud, silt or sediment
+ **Rarapa:** glistening

+ **Rau:** many or a hundred
+ **Roa:** long
+ **Rua:** pit
+ **Roto:** lake, inside
+ **Tai:** coast, tide
+ **Taki:** noisy or weeping
+ **Tere:** swift
+ **Tiri:** to toss about or disturb
+ **Tūtae:** excrement
+ **Wai:** water
+ **Whanga:** harbor, bay

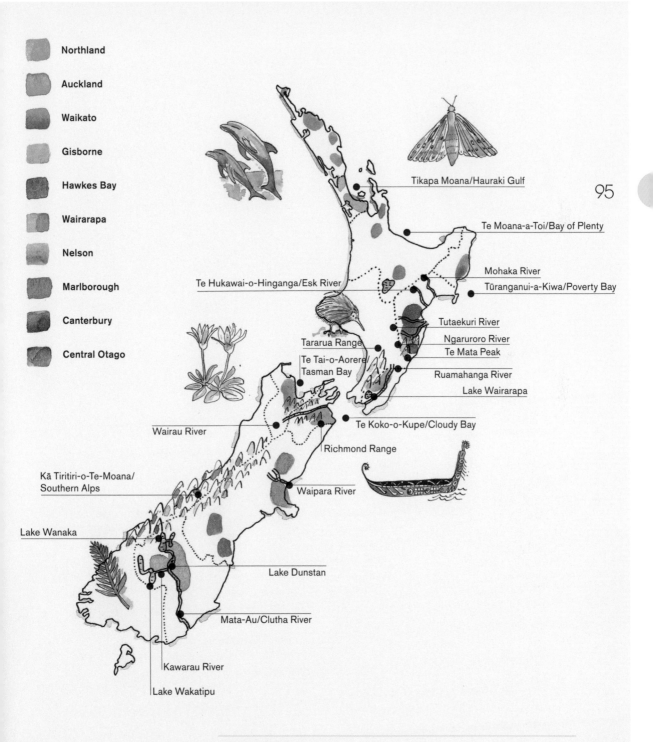

Northland
Auckland
Waikato
Gisborne
Hawkes Bay
Wairarapa
Nelson
Marlborough
Canterbury
Central Otago

Tikapa Moana/Hauraki Gulf

Te Moana-a-Toi/Bay of Plenty

Mohaka River

Te Hukawai-o-Hinganga/Esk River

Tūranganui-a-Kiwa/Poverty Bay

Tutaekuri River

Ngaruroro River

Te Mata Peak

Tararua Range

Te Tai-o-Aorere/
Tasman Bay

Ruamahanga River

Lake Wairarapa

Wairau River

Te Koko-o-Kupe/Cloudy Bay

Richmond Range

Kā Tiritiri-o-Te-Moana/
Southern Alps

Waipara River

Lake Wanaka

Lake Dunstan

Mata-Au/Clutha River

Kawarau River

Lake Wakatipu

Gin

96

Gin in a Negroni, gin with tonic,
and a gin martini are each reserved
for very different moods.

I was having a Negroni and planning out my week in a restaurant near my new apartment. Jon was on my mind. I emailed him – strictly business, I told myself – about a tasting group the following Monday. He responded promptly with my requested information, but he also asked me how I was. We shot back and forth a few niceties. He told me he missed me. I told him I missed him. He asked if he could come over that night. I said yes.

I told myself I had to be strong. The way he talked to me and the way he treated me were unacceptable. *I am worthy of someone who adores me and dotes on me. I am not a victim.* I felt weak to let him come over at all, but just to be in his presence and feel his touch made me feel more like me. He brought a bottle of Champagne. A house-warming present. He said he was sorry for the other night.

"Did you mean what you said? That you could never fall in love with me?"

"No. It just seemed like you wanted everything at once."

"No. I just want to be with you."

We kissed. He spent another restless night. I was unconvinced of his dedication, but so utterly convinced of my own feelings.

We made plans to go out for a nice dinner on our next day off.

VIGNETTE

I made the reservation: a trendy restaurant in SoHo, known for its intimate setting and bombastic table-side preparations of throwback American-Italian food. I wore a new white dress, and too much perfume, trying to mask the scent of a new deodorant that smelled like gin and tonic with a three-day-old lime wedge.

I got to the restaurant before he did, checked in with the hostesses, and ordered a drink at the standing-only bar. It was a Last Word, a potent but refreshing combination of gin, Chartreuse, lime, and maraschino liqueur. Fifteen minutes went by. I nursed my drink and shifted my weight. I checked my phone to see if maybe I'd missed a text from him. Nothing. I texted him. The text remained undelivered, like his phone was off. Twenty-five minutes after our reservation time, and still no word. I went outside to call him. Straight to voicemail.

I smiled curtly at the hostesses when I came back in. I explained, as unemotionally as I could, that I had no idea where my date was. "So odd!" I said, as if I had misplaced my ID while being carded underage. "No clue!" I said, as if I would find it any minute in the pocket of my purse. "I'll wait for just a few minutes longer."

They were sweet and smiled back and told me it was no trouble.

I spent a few more minutes fidgeting at the bar before paying for my drink. I walked to the host desk. "I guess I'll throw in the towel!" I said, smiling. I apologized for taking up the table. They assured me not to worry, a mix of pity and amusement in their eyes that I could not blame them for.

I walked north, passing the trendy shops and barred-up bodegas of western SoHo. I called Jon again, fuming, furious, and on the verge of tears. This time, the phone rang, but no answer. I waited a few minutes and called him again. I was on the southern border of Washington Square Park, way too dressed up next to the NYU kids and stoned fountain-loungers who lined the streets. He picked up. His voice was raspy, and he sounded like he'd just woken up.

"Are you okay?"

He was wailing, almost crying.

Finally: "I'm so drunk right now. Am I supposed to be somewhere?"

I could barely speak. Tears fell out of my eyes and onto the pavement. "Yeah, we had a DATE, Jon!"

"No, no, nooooo, no, nooo…"

He seemed legitimately upset. I wanted so badly for him to give me something by which I could forgive him. At that point, I would have taken almost anything. I waited in an excruciating moment of silence, wanting to hang up, but also needing so much more from him. *What was he doing? How could he do this to me? And – why?* Still, nothing.

So I hung up. I walked for close to an hour, finding myself in Gramercy. New York is an amazing city when you're sad. The buildings stooped and trees bowed with sympathy. Every passer-by seemed to have a glimpse of sadness in their eyes, telling me *you're not alone*. When I finally landed at a bar that night, I ordered a gin martini: a drink I reserved exclusively for moments where I felt a desperate sense of longing.

Prohibition:
A Spirits Guessing Game

Stretch your mind and your vocabulary by trying to get your teammates to guess the title word on the card, but without saying any of the words listed below it. Choose your level of difficulty:

1. Any word not on the card is fair game.
2. No specific name-brands may be used.
3. No specific name-brands or cocktail names may be used.

GIN

Juniper
Tonic
Negroni
Genever
Beefeater
Tanqueray

JAPANESE WHISKY

Scotch
Whisky
Single Malt
Barley
Yamazaki
Hibiki

SCOTCH

Scotland
Whisky
Single Malt
Barley
Speyside
Islay

BRANDY

Grape
Fruit
Apple
Cognac
Armagnac
Sidecar

TEQUILA

Agave
Mexico
Jalisco
Reposado
Margarita
Patrón

BOURBON

America/USA
Rye
Corn
Whiskey
Kentucky
Jim Beam

VODKA

Neutral
Grain
Cosmo
Soda
Belvedere
Stoli

COGNAC

Brandy
Grape
France
Sidecar
Hennessy
Rémy Martin

MEZCAL

Tequila
Agave
Mexico
Oaxaca
Del Maguey
Smoky

RYE

America/USA
Bourbon
Oak
Whiskey
Kentucky
Manhattan

GENEVER

Gin
Juniper
Malt
Holland
Dutch
Bols

GRAPPA

Grape
Brandy
Pomace
Italy
Romano Levi
Nonino

CHARTREUSE

France
Alps
Monks
Yellow
Green
Carthusian

CALVADOS

France
Apple
Pear
Pays d'Auge
Normandy
Dupont

RUM

Sugar
Molasses
Caribbean
Central America
Bacardi
Daiquiri

RHUM AGRICOLE

Cane Sugar
Martinique
West Indies
France
La Favorite
Ti Punch

EAU-DE-VIE

Fruit
Brandy
Water of Life
Unaged
Plum
Raspberry

IRISH WHISKEY

Grain
Unmalted Barley
Pot-Stilled
Jameson
Red Breast
Dublin

DOMAINE HUET *LE HAUT-LIEU SEC* Vouvray, Loire Valley, France
NICOLAS JOLY *LES VIEUX CLOS* Savennières, Loire Valley, France
JEAN-PIERRE ROBINOT *JULIETTE* France

Chenin Blanc

I never dated much growing up.

I was taller than all the boys in high school, had frizzy hair, acne, and braces. Perhaps I could have attracted someone, but I certainly didn't have the confidence. So I studied and dreamed and became a person I liked. I didn't need a man to define my worth or make me happy. If anything, the whole endeavor stressed me out. I didn't understand my girlfriends who darted from boy to boy as if they were shelters in a minefield. I was my own shelter, *thankyouverymuch*.

This attitude made my feelings for Jon all the more confusing. Even after that night, I continued to crave him. A few days later, I texted him to make sure he was okay. He texted me back. Yes, he was okay. He apologized, telling me I didn't deserve what had happened. He'd had a long afternoon of drinking with friends and had simply fallen asleep. An explanation, but no true remorse. He never made any movement to atone or seek forgiveness. He was content to let me slip away. And my pride wouldn't let me pursue a man who had made it so clear – in actions and in words – that I was unimportant to him.

Over the summer of 2013, Jon and I continued to see each other in professional settings. One day, a morning tasting group led to a noon lunch together, which led to an afternoon in my bed. I told myself it was just physical and that I could handle it. We continued to see each other in this capacity from time to time. Late nights and cold mornings, with no real affection. It left me sad and

---

CHENIN BLANC

hollow, but it seemed better than the alternative: cutting him out of my life altogether.

That summer, I was also wait-listed for the Court of Master Sommeliers Advanced exam that would take place in August of 2013. I was bummed to be wait-listed. I didn't want to wait another year. I was ready.

Jon texted me one afternoon. His name popping up on my phone always sent butterflies through my stomach and a smile to my lips. "Check your email," he wrote. "Caleb just got off the wait-list, maybe you did too." I checked my email. I had indeed gotten a seat at the August Advanced exam. I told Jon the news. "You got this," he replied.

I was at a wine store during this exchange, buying for a tasting group. I had to step outside as tears flooded my eyes. I couldn't exactly explain why I was crying. My emotions over Jon and over the exam collided, and I couldn't distinguish the happy tears from the sad ones.

I went back inside the wine store and bought a few different bottles of Chenin Blanc. It was a grape I had been struggling with in blind-tastings because its personalities are so diverse and often defy obvious classification. You are always supposed to deduce in the blind-tasting format, but Chenin is one of those grapes that defies deduction. Either it feels like Chenin, or it doesn't.

Part of Chenin's unique appeal is that its flavor profile is hard to replicate outside of the Loire Valley. There are some nice bottlings of Chenin Blanc in South Africa, Australia, California, and New Zealand; but in general, the grape is most itself in the 150 kilometers between Vouvray and Savennières. In that small tract of land, it transforms from fiercely lean and dry to opulent, exotic, and spiced.

Chenin Blanc is also a bit of an underdog. Its prices have never reached the realm of Chardonnay from Burgundy, and its lack of universal planting means its name is never on the tip of the masses' tongues. Its expression, though, will always endure: its sense of singularity that defies explanation, and its ability to be felt without being thought about.

Chenin Blanc's Spirit Humans

When I first got into wine, I was asked: what is the wine that best describes your personality and why? My reply was Chenin Blanc: "Can be cheerful and lighthearted; can be intense and dramatic. Never at home at a football game."

The anthropomorphism of wine can be obnoxious, but Chenin Blanc truly does seem to have more personality (in the very human sense) than other grapes. Explore the wild, wonderful, and beautiful styles of Chenin through these different genres of twenty-something women, as typified by famous twenty-somethings.

Vouvray

Emma Watson: classic, restrained, intellectual

Saumur

Jennifer Lawrence: regal, structured, powerful

Anjou

Taylor Swift: unfettered, wild, dramatic

Savennières

Miley Cyrus: unpredictable, fiery, outgoing

Quarts de Chaume

Rihanna: outspoken, majestic, accomplished

Coteaux du Layon

Margot Robbie: wise, playful, warm

DOMAINE DE CHEVALIER Pessac-Léognan, Graves, Bordeaux, France
DOG POINT VINEYARD *SAUVIGNON BLANC* Marlborough, New Zealand
FRANÇOIS COTAT *LES MONTS DAMNÉS* Sancerre, Loire Valley, France

Blind-Tasting Wine

Blind-tasting requires built skills
as well as the ability to refrain from
"playing the game."

Sommeliers like to think about what the wine might be, should be, and could be, when we need to be thinking about what the wine actually *is*.

I had regularly been attending the blind-tasting group at Eleven Madison Park. One of our favorite formats was for each person to bring two wines, a red and a white. Everyone would partner up, then the first person in each partnership would pour their wines to create a flight of six. The partner who didn't pour would taste, and then pour their wines in the second flight. This allowed for flights of six to be created without any one person having to bring all six wines. The downside was that sometimes the flights weren't very well balanced. For instance, a number of times all three red wines, say, were from the same country. A few times, the exact same wine was replicated. And once, to my knowledge, all three white wines were the same grape.

This happened one of the first times I tasted with the EMP group. I was the first to taste, my partner having poured his wines in the flight. I started in with the first white wine. I was less methodical and more instinctual at this point in my tasting career. I got ruby-red grapefruit, passionfruit, charred green pepper, and gooseberry. It was ripe but fresh, exuberant in its fruit and green characteristics. *New Zealand Sauvignon Blanc, done.* I moved on to the next wine. Still that green and tropical character, but creamier, nuttier, oakier. The green was more celery salt than green pepper, and the fruit was more lemon curd and plantain chip than ruby-red grapefruit and gooseberry. *Bordeaux Blanc, next.* The third wine was a paired down version of the first two. Subtle white grapefruit, clean grassy notes, and burnt-match flintiness. It was like the first two taken off steroids. *Sancerre.*

I don't remember the reds in that flight, and I probably didn't get them all right. But the white wines I had nailed, and I was the only one to do it. I was in the infancy of my blind-tasting career, and something about my naïveté allowed me to not question three Sauvignon Blancs in a row. Everyone else had figured they were making things up or allowing the wines to influence one another. But I didn't think twice about it.

Though this was a great blind-tasting performance, it would be years before I became a great blind-taster. It is a skill that takes a long time to develop. But developing blind-tasting skills is one of the most important things we do as sommeliers. A lot of people assume it is essentially a party trick, but it is quite the opposite. It is the foundation of our trade.

As sommeliers and wine buyers, we need to be able to asses quality in order to put something on our list. *Is this a sound, clean wine? Is it well made? Does it represent value for its price point?* We also need to understand how a wine falls in the spectrum of its style in order to describe it to a guest. *Is this wine abnormally oaky for its style? Ultra-concentrated? Lighter and leaner than usual?* The only way to know these things is to know what classic wines taste like, why, and then to be able to taste those qualities blind.

If I have two Sancerres on my list, and one is razor-sharp and very clean, and the other has a bit of botrytis and oak texture – this is important information for the guest to know. And it's information that may not be available in professional reviews or on winery websites. It will only be available to those who can taste it. And the only people who will be able to taste it are the ones who practice blind-tasting.

VIGNETTE

Blind-Tasting Flow Charts

Tasting wine can be quite complex: is that apricot I'm smelling or nectarine? Is the **lees** quality more like parmesan cheese rind or stale beer? Is that rutabaga or white radish? Is this Barolo or Barberesco? Is this Chambolle Musigny or Vosne Romanée? The nuances are endless and can be difficult to parse.

Blind-tasting for an exam, though, is simple. This doesn't mean it is easy, but it is uncomplicated. It is about taking a wine and distilling it to the three to five most important qualities: the features that make it like nothing else in the world.

The following charts for red and white wine track my split-second thinking when I'm in the thick of an exam or flight. Each colored splash represents a starting place: if I can recognize this particular aroma (and I usually can), I can trust my logic-tree to get me to the right conclusion. If I *can't* put my finger on one of those starting points, I know I can start with something else – **residual sugar** (RS), **volatile acidity** (VA), alcohol, **tannin**, color – and I can muscle my way back to the start.

It is imperative for students of blind-tasting to develop their own vocabulary, understand their own logic, and grow to trust their process. Each palate is different, and what might lead me to the right conclusion may not work the same for someone else. Use my charts as examples to build your own, rather than models to dogmatically follow.

Blind–Tasting Flow Charts

Red Wine

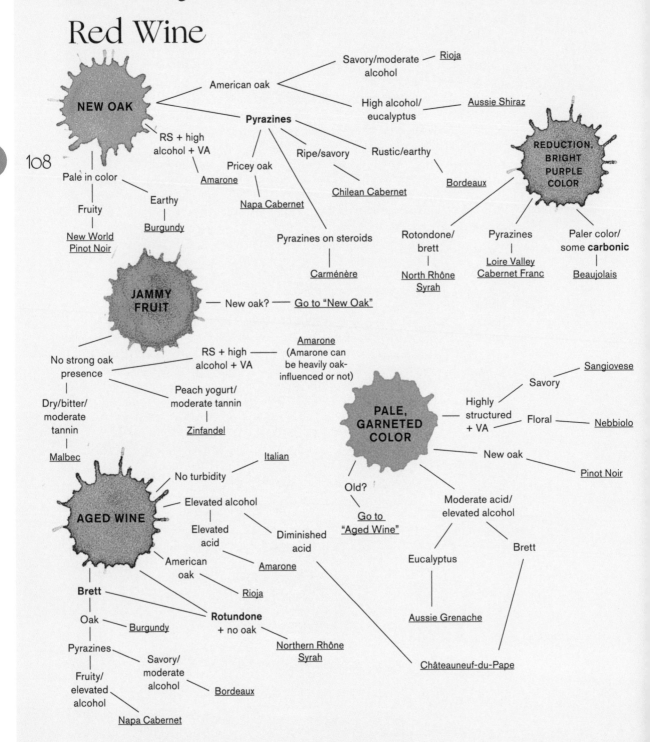

White Wine

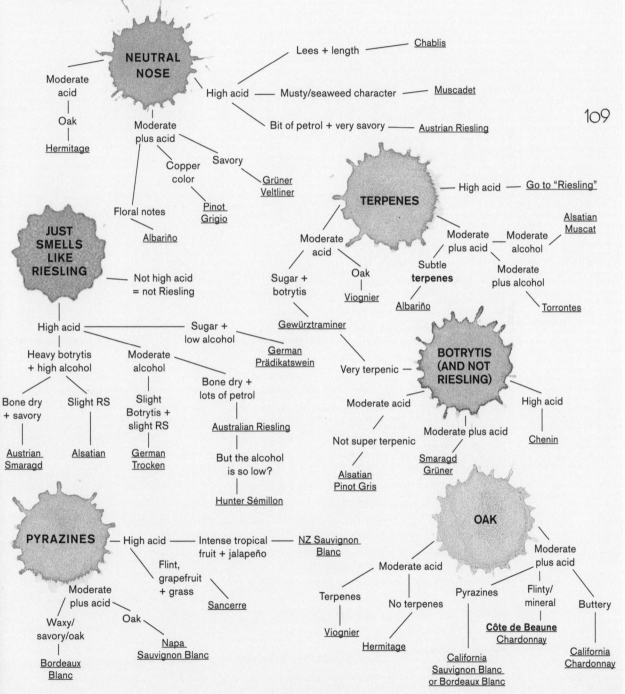

NEUTRAL NOSE

Lees + length — Chablis

High acid — Musty/seaweed character — Muscadet

Bit of petrol + very savory — Austrian Riesling

Moderate acid — Oak — Hermitage

Moderate plus acid

Savory — Grüner Veltliner

Copper color — Pinot Grigio

Floral notes — Albariño

JUST SMELLS LIKE RIESLING

Not high acid = not Riesling

TERPENES

High acid — Go to "Riesling"

Moderate plus acid — Moderate alcohol — Alsatian Muscat

Subtle terpenes — Albariño

Moderate plus alcohol — Torrontes

Moderate acid

Sugar + botrytis — Gewürztraminer

Oak — Viognier

High acid

Sugar + low alcohol — German Prädikatswein

Very terpenic

Heavy botrytis + high alcohol

Bone dry + savory — Austrian Smaragd

Slight RS — Alsatian

Moderate alcohol

Slight Botrytis + slight RS — German Trocken

Bone dry + lots of petrol — Australian Riesling

But the alcohol is so low? — Hunter Sémillon

BOTRYTIS (AND NOT RIESLING)

Moderate acid

Not super terpenic — Alsatian Pinot Gris

Moderate plus acid — Smaragd Grüner

High acid — Chenin

PYRAZINES

High acid — Intense tropical fruit + jalapeño — NZ Sauvignon Blanc

Flint, grapefruit + grass — Sancerre

Moderate plus acid

Waxy/savory/oak — Bordeaux Blanc

Oak — Napa Sauvignon Blanc

OAK

Moderate acid

Terpenes — Viognier

No terpenes — Hermitage

Moderate plus acid

Pyrazines — California Sauvignon Blanc or Bordeaux Blanc

Flinty/mineral — **Côte de Beaune** Chardonnay

Buttery — California Chardonnay

109

American Whiskey

I was pumped for the Advanced exam.

I had been feeling good, sleeping well, and I had quit my restaurant bar job about a month before. I was studying a lot, tasting a lot, and carrying a tray full of way too many Champagne flutes around our living room. I didn't leave anything to chance. I mapped out exactly what my **gueridon** would look like during the practical portion. I created a custom wine list that I could draw upon when asked for pairing suggestions. On the plane flight over, the other New York candidates and I threw theory questions back and forth, all of which I knew.

The one section I was a bit nervous about was tasting. My last flight before the exam was not my greatest (I called Chablis as Pinot Grigio and Malbec as Saint-Joseph). I had shipped a six-pack of wine to the hotel in advance of the exam. A few wines that had been giving me trouble, just to re-familiarize myself throughout the week, plus a bottle of Fernet-Branca and a bottle of Eagle Rare 10-year-old. Since The Violet Hour days, American whiskey had always been my go-to nip to chill out.

In addition to my "problem wines" showing up during tasting, I was worried about getting stuffed up. Sommeliers, especially just before an exam, take an abnormal interest in their sinus health. I'm sure every candidate had a neti pot and a copious supply of Sudafed, and I was no different.

I took the Advanced exam the last year in the US that the course and the exam were in the same week. Now, candidates who want to sit for the Advanced attend a separate three-day course beforehand. This is probably for the best, as most of us were so anxious for the exam that we didn't get too much out of the course.

On the final day of the course – the day before the exam – I was complaining about feeling stuffed up (my way of saying, *if I don't pass, this is the reason*). A Canadian sommelier offered me some Sudafed.

"Oh, I have some. It's just not doing much. Thank you."

"Do you just have over-the-counter stuff?"

"Well, yes…"

"This is the real stuff," he said, palming a few pill tabs toward me. I had never taken part in a drug deal before, but I imagined this was the way it went down.

"Okay, well, yeah, that sounds good," I said, taking the pills from him. I didn't know much about the difference between "real" and "fake" Sudafed. All I knew was that you had to give your ID to the pharmacy if you wanted the "real" stuff because you could somehow turn it into crystal meth (which I didn't know anything about either; this was before *Breaking Bad* was a thing). But, I figured, the "real" stuff had to work better than my over-the-counter brand.

I took two pills and went back for the afternoon session. Within an hour I was in the throes of what can best be described as a paralyzed panic attack. My heart was throbbing in my chest, my limbs were ringing, but I could hardly keep my eyes open. I didn't make the connection at first. I thought my nerves were just finally kicking in. I struggled through the rest of the afternoon's programming – a few lectures that I didn't retain a word of, and a painful blind-tasting session where I called Rioja as Châteauneuf-du-Pape. I was frantic over my new symptoms. I also panicked about the Masters seeing me like this, and thinking I was being rude by dozing off during lectures.

I went back to my hotel room and took a shower. I barely got myself dried off and put some underwear on before I collapsed on my bed. When I woke up the sun was setting. The throbbing in my chest was even stronger than before. My whole body felt like it had been set on fire from the inside. I tried to reach for my phone to check the time but my hand would not move. I stared down at my left hand. I willed myself to lift it, and my right pointer finger barely came off the bed.

Okay, that's a start, I thought. I knew my options were either to be positive or start sobbing.

I looked down again and got my middle finger to lift. Then my ring finger. Then my pinky. My thumb took a few more minutes. I finally got my hand off the bed and lobbed it toward my phone on the nightstand. I called my mom. At this point, I couldn't hold back the tears. It wasn't so much sobbing as silent streams falling from each eye. All the work I had put into this – both into my health and into the exam – seemed to be slipping away by the second.

My mom told me to call the hotel. I did. Within a few minutes a hotel manager was in my room. I must have been a sight: bra and underwear, splayed out on the bed, wet hair, face red and tear-stained. She asked me how I was doing and if I had consumed anything unusual that day.

"Yeah, I took a few tablets of Sudafed. You know, the real stuff." I'm sure she was amused by my description.

"How do you normally tolerate stimulants?"

Ohh. Finally, it all made sense. "Umm. I don't," I sighed. "I don't drink coffee or even green tea. They make me tired and panicked."

"Okay," she said, "I'm still going to send you to the hospital, for them to take a look."

Within ten minutes, several EMT were in my room. By this point, the manager had at least covered me with a blanket. I still couldn't move very easily so they lifted me onto a gurney. I was rolled through the hotel; a few candidates and Masters definitely witnessed the scene. Tears came to my eyes again as I remembered I had to start the exam the next day.

I spent the next few hours in a hospital bed. I had my phone and was able to text my parents updates. I also texted a few friends who were taking the exam to let them know where I was. One of them emailed Shayn, the examination director, to keep him in the loop. I knew this was the right thing to do, but I didn't want it to become a big deal. I also texted Jon, who didn't seem very concerned. "This will be a great story when you pass," he said.

Within a few hours, I had regained full mobility. My nervous system had calmed down. I could keep my eyes open. The doctors attributed the reaction to a combination of the Sudafed, the thyroid medication I was still on, the Klonopin, and probably the stress of the week. They said if I was feeling better, I didn't have to spend the night.

I got back to my hotel room around 11pm. I had to be up at 8am for another lecture, then the written theory exam in the afternoon. I was still pretty wired from the day. I poured myself a splash of the Eagle Rare and closed my notebooks. *No more studying, no more worrying. You are prepared. You can do this, Jane.*

I dozed off quickly, and the whiskey lulled me into a peaceful sleep. I woke up the next day feeling a little drowsy, but overall just fine. When I walked into the morning lecture, just about every MS asked me how I was doing, like I was going to break. "I'm fine, thank you," I said, smiling, on repeat.

It was kind of them, but I did not want anyone's pity. I also didn't know what story was going around. My friend had told Shayn that I'd had an allergic reaction, but I was being treated like I'd had a mental breakdown. Or at least that's what it felt like (which could have been my projection). I knew I had to put it out of my mind and focus on the next few days.

The exam portions of both the Introductory and Certified levels take place in one day. The stamina and mental fortitude required for a multiple-day exam is a whole different beast. I ran on the hotel gym treadmill every day, sometimes a few times. I took baths. I went out to dinner with friends. I tried to balance relaxation with distraction. I had brought all my study materials, but it seemed pointless once I got there. Cramming felt depleting, not enhancing.

The time between the sections felt the hardest. The exam itself was smart, fair, well-written, and – surprisingly – quite fun. After our New York contingent had finished the final section, we convened in my hotel room and drank Fernet-Branca and Eagle Rare. It was early afternoon and we had to make sure we were composed for the results that evening, but it felt good to let loose after all the stress.

Laura Maniec gave me my results. Laura is a New York-based Master Sommelier and one of the best businesspeople in the industry. She was heavily involved in the Court and we had met a few times before the exam.

"Theory was good, very strong. You passed that."

I nodded, smiled, and took a deep breath.

"You had some good tables in service," she said, shuffling through the notes. "Overall a very strong performance there. You passed service."

I held my breath.

"Tasting, there were a few issues with some structure calls, and you sped through a few notes." I looked down, prepared to

---

AMERICAN WHISKEY

hear the worst. "But overall quite good. You passed that as well. Which means…"

She looked at me and smiled, and I smiled back. "You passed!"

She gave me a hug and walked me back to the hall where everyone was waiting. My friends – some of whom passed and some didn't – greeted me with open arms. Masters shook my hand and texts and emails flooded in from family and friends.

I knew at that moment that I wanted to be a Master Sommelier. I wanted to put in the hard work to continue to be a part of this community. I wanted to be in Laura's position some day: giving someone else the news that their hard work had paid off.

We drank Champagne at the reception, then more wine at dinner. The exact bottles I can't remember. When I finally got back to the hotel room, there was about a finger left of the Eagle Rare. I did not need any more booze, but I felt like I needed to bookend. I thought of the last time I drank it, just after finishing the exam, but before knowing the results. And I thought of the time before that, the night before the exam, when I got back from the hospital. Each of those moments was full of doubt and fear. But each was also quite beautiful, in its own way (and especially in hindsight), and the doubt and fear made the moment of passing all the sweeter.

I took the last swig out of the bottle and collapsed into bed.

A Family Tree of American Whiskey

The production of American Whiskey defies the standards of the spirit (and wine) world. While most spirits and wine are produced at a distillery (or winery) that bears its name, nearly all the mainstream American whiskey brands are made at only a handful of distilleries.

In *The Kings County Distillery Guide to Urban Moonshining*, Colin Spoelman and David Haskell present the most useful graphic I've found to understand the major brands of "Kentucky whiskey" (their term for American whiskey made by large corporations in and around Kentucky). I've included a version of it on the next pages.

Looking at this chart begs the question: what distinguishes one whiskey from another made at the same distillery? They all come off the same still with a limited number of mash bills. This leaves us with age statement, **proof**, and barrel selection as the distinguishing factors.

For wine, we don't look at the ABV on the bottle and the time spent in oak as measures of quality. We as sommeliers primarily look at the raw material – the grapes – in order to make a qualitative judgement about the wine. Where did they come from? How were they grown? We also talk about the ingenuity of the vinification process (the human factor) and, most important, how it actually tastes!

Without the shackles of legal requirements and production quotas, craft distillers have brought these factors to the whiskey conversation and imbued the industry with creativity. The result is a landscape of different grains in American whiskey, unique stills, and experimental production and aging.

Unfortunately, because of the age-statement frenzy that Kentucky whiskey created, many craft distillers feel compelled to compete on those terms and purchase aged whiskey from one of the big distilleries to market as their own. Similarly, Kentucky whiskey wants to capitalize on the goodwill generated by the craft movement, and thus litters its labels with stories of founding pioneers and Lilliputian outputs.

I believe each sector should do what it does well and stop trying to compete. Craft distilling should focus on the creativity of production and the quality of its raw materials: factors difficult on a mass level. And "Kentucky whiskey" should focus on turning out a consistent and well-aged product at a good price: a hard feat for a small or fledgling distillery. There is no qualitative superiority in a high age statement. And there is no moral superiority in a small batch.

A Family Tree of American Whiskey

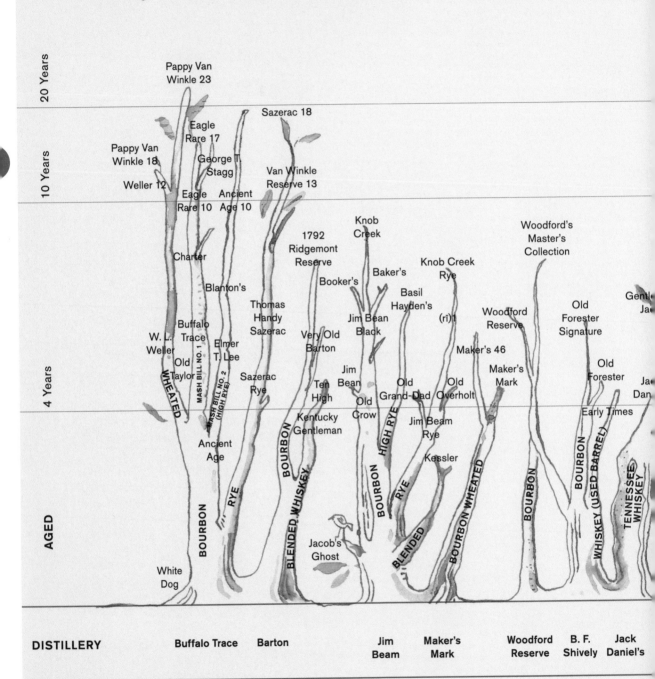

20 Years

Pappy Van Winkle 23

Sazerac 18

10 Years

Eagle Rare 17

Pappy Van Winkle 18

George T. Stagg

Van Winkle Reserve 13

Weller 12

Eagle Rare 10 Ancient Age 10

Knob Creek

Woodford's Master's Collection

Charter

1792 Ridgemont Reserve

Knob Creek Rye

Baker's

Blanton's

Booker's

Basil Hayden's

Gentl Ja

Thomas Handy Sazerac

Jim Beam Black

(rī) 1

Woodford Reserve

Old Forester Signature

4 Years

Buffalo Trace

Very Old Barton

Maker's 46

Old Forester

W. L. Weller

Elmer T. Lee

Old Taylor

Jim Beam

Old Grand-Dad

Old Overholt

Maker's Mark

Ja Dan

WHEATED

MASH BILL NO. 1

MASH BILL NO. 2 (HIGH RYE)

Sazerac Rye

Ten High

Old Crow

Early Times

Kentucky Gentleman

Jim Beam Rye

Ancient Age

BOURBON

RYE

BOURBON

BLENDED WHISKEY

HIGH RYE

Kessler

BOURBON

RYE

BOURBON WHEATED

BOURBON

BOURBON

WHISKEY (USED BARREL)

TENNESSEE WHISKEY

AGED

BLENDED

Jacob's Ghost

White Dog

DISTILLERY

Buffalo Trace   Barton   Jim Beam   Maker's Mark   Woodford Reserve   B. F. Shively   Jack Daniel's

CORPORATE OWNER

Sazerac   Suntory   Brown-Forman

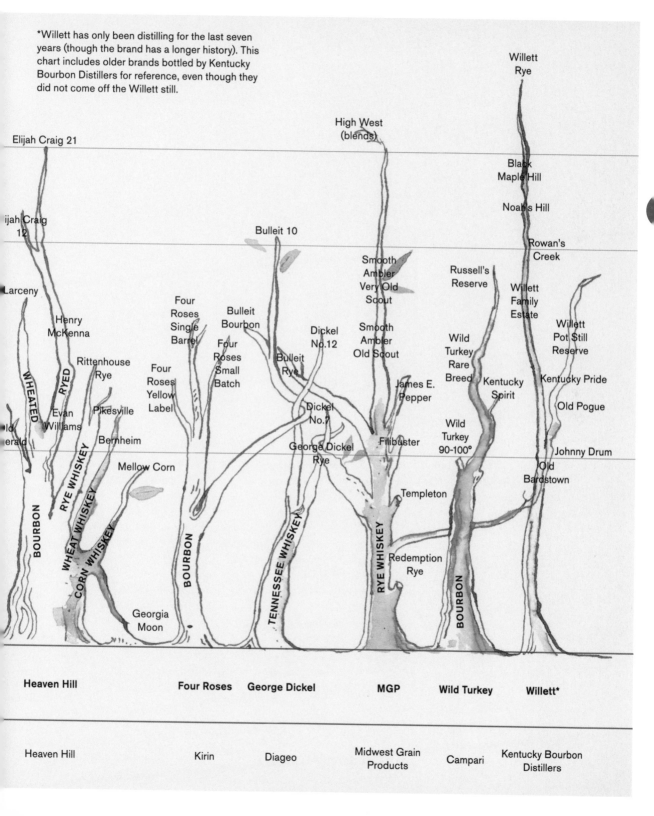

*Willett has only been distilling for the last seven years (though the brand has a longer history). This chart includes older brands bottled by Kentucky Bourbon Distillers for reference, even though they did not come off the Willett still.

Willett Rye

High West (blends)

Elijah Craig 21

Black Maple Hill

Noah's Hill

ijah Craig 12

Bulleit 10

Rowan's Creek

Smooth Ambler Very Old Scout

Russell's Reserve

Larceny

Willett Family Estate

Four Roses Single Barrel

Bulleit Bourbon

Dickel No.12

Smooth Ambler Old Scout

Willett Pot Still Reserve

Henry McKenna

Rittenhouse Rye

Four Roses Small Batch

Wild Turkey Rare Breed

Kentucky Pride

Four Roses Yellow Label

Bulleit Rye

James E. Pepper

Kentucky Spirit

RYED

Old Pogue

WHEATED

Evan Williams

Pikesville

Dickel No.7

Old erald

Bernheim

George Dickel Rye

Filibuster

Wild Turkey 90-100°

Johnny Drum

Mellow Corn

Old Bardstown

RYE WHISKEY

Templeton

WHEAT WHISKEY

BOURBON

CORN WHISKEY

BOURBON

TENNESSEE WHISKEY

Redemption Rye

RYE WHISKEY

BOURBON

Georgia Moon

Heaven Hill   **Four Roses   George Dickel   MGP   Wild Turkey   Willett***

Heaven Hill   Kirin   Diageo   Midwest Grain Products   Campari   Kentucky Bourbon Distillers

WEINGUT EGON MÜLLER *SCHARZHOFBERGER RIESLING KABINETT*
Mosel, Germany
WEINGUT LEITZ *RÜDESHEIMER BERG SCHLOSSBERG* Rheingau, Germany

German Riesling

It was mid–August in Manhattan.
I was an Advanced sommelier and jobless.

I had a sommelier gig lined up with a new restaurant that was supposed to open in early October. I had a paid trip to Alto Adige for a wine trip there in September, so I decided to extend the trip and explore Germany before and more of Italy after. I would spend two weeks, alone, on the road, split by one week with other sommeliers in the middle. I can't recommend the solo wine road-trip enough. These are the lessons I took from it.

Lesson #1
Look at a map.

I set general benchmarks that, in hindsight, made very little sense. I flew into Frankfurt and started in the Pfalz. From there I drove down to Alsace, then back up to Nahe. Then onto Rheinhessen, Rheingau, and finally the Mosel. I darted from one region to the next, crossing country lines, like everything was right next door. It wasn't. It took a long time, and I was almost late to several appointments.

Lesson #2
Get a great GPS.
And a SIM card.

In 2013, I wasn't clued in to the joys of SIM cards. I kept my American phone number and only had internet when I was connected to Wi-Fi. This all sounds very romantic, a peaceful disconnect from the e-world, until you are dead lost in the wilds of Italy.

Germany has the GPS thing down pat. My elite, leather-seated German rental car had a perfect English-speaking GPS. She saved me time. She kept me company. We laughed. We cried.

Italy was a different story. My GPS was robotic in tone and could never figure out where I was. "Turn left" became "turn right" became "go back" all while I was sitting at a stoplight. She finally gave out about halfway through my trip, and I racked up hundreds of dollars in roaming fees letting my iPad guide me. But what was the alternative? Use a map? This wasn't the 1980s!

119

Lesson #3
Just ask.

I emailed the wineries I wanted to see, asking if I could make an appointment to visit them. I explained that I was a sommelier in New York, which I thought would make me sound more legitimate, but probably just made me sound like more of a tool. (Especially because, well, I wasn't a sommelier in New York. I had never been a sommelier in New York. I was unemployed with a shiny pin.)

Every winery but one responded with open arms and treated me like a VIP. I was welcomed into the Faller's living room for tea at Domaine Weinbach. Jean Trimbach poured me 1990 Cuvée Frederic Emile blind. Johannes Leitz invited me to a party and I stayed up all night drinking old Berg Schlossberg with him and his friends. Helmut Dönnhoff even suggested we take a picture together at the end of my visit, about which I am still fangirling.

And all I did was ask. I was respectful. I was polite. I showed up on time. And I took genuine interest in the special things these people brought to the world.

Lesson #4
Plan some,
but not too much.

I booked a hotel room in each town. I had a few visits planned. But I left time to wander and explore. And, most importantly, I didn't book any restaurants beforehand. At each appointment, I simply asked my host where I should eat that night. Winemakers ALWAYS know the best places to eat and drink. I was turned on to every kind of wonderful establishment: from the schnitzel-and-carafe-of-Riesling joints to the ten-course-tasting-menu-of-your-dreams. Plus, it never hurts to say in broken German/French/Italian that "[insert great winemaker] sent me here."

Lesson #5
Do something you wouldn't normally do.

BE SAFE, okay? But maybe take advantage of being alone and out of your comfort zone to find an unusual activity. I found one on a Sunday in the Nahe. Everything (and I mean EVERYTHING) was closed. No wineries were taking visits; no restaurants were open. I literally googled "what to do in the Nahe on a Sunday" and found an emphatic TripAdvisor page for something termed a "textile-free spa." *Hmm…*I thought. *Like, they don't have towels?*

I read on and soon came to understand what it meant. *It's a nude spa!* It looked gorgeous, though: stone saunas, open-air baths, lush rooftop gardens. And in every picture, naked German people were having a great time. *You know*, I thought, *why not? I won't know a soul. It might be sort of liberating.*

I checked in at the front desk and was given a locker and a robe by – no joke – a woman with the name tag "Brünnhilde." I was told that I could wear the robe around, but that it wasn't allowed in the saunas or steam rooms. "Nothing is."

"Nothing is…what?" I asked.

"Nothing is allowed in the rooms." She could tell I still didn't get it. "No clothing," she said, pointing almost disdainfully at my garments.

"Yeah, okay, that's fine, *ja*," I said, nodding effusively. It was less of a consent to their rules and more of an attempt to convince myself that I could do it.

I changed in the locker room. I'd mastered that wiggle maneuver you do in middle school gym class, where you take off your bra and underwear without removing your towel (or, in this case, robe). Even though I was alone in the room, I still felt odd to be naked in public.

I walked down to the spa door and casually slipped my robe onto the hook outside. *I do this all the time* was the vibe I was going for, but I don't think I could've fooled anyone. When I walked inside, much to my delight, it was completely empty. I sat down on a stone bench, which was very hot on my bare bum, and finally started to relax. Soon, though, the door opened. A man walked in and sat across from me. He was probably in his fifties. He was squat and soft and had more hair on his chest than on his head. And he was very, VERY naked. He splayed his legs out wide, threw his back up against the rest, crossed his arms on the ridge created by his belly, and closed his eyes.

He did not give a shit about me. I soon realized that no one did. This wasn't a place to leer, or gawk, or judge. Everyone was just there to relax, enjoy the space, and – literally and figuratively – let it all hang out.

120

VIGNETTE

Decoding a German Wine Label

German wine labels can be some of the most intimidating to stare down. These annotated labels can help you suss out style, region, and quality level when you're next in the Riesling aisle. Most importantly, they can help you decode how the wine will taste – namely, will it be sweet or dry? There are other important terms that are not on the following labels, like **Grosses Gewächs**, **feinherb**, and **sekt**. Head to the glossary for these definitions.

Vintage: The year the grapes were picked. This one's pretty universal.

Anbaugebiete: The major growing region of production. There are 13 such areas in Germany.

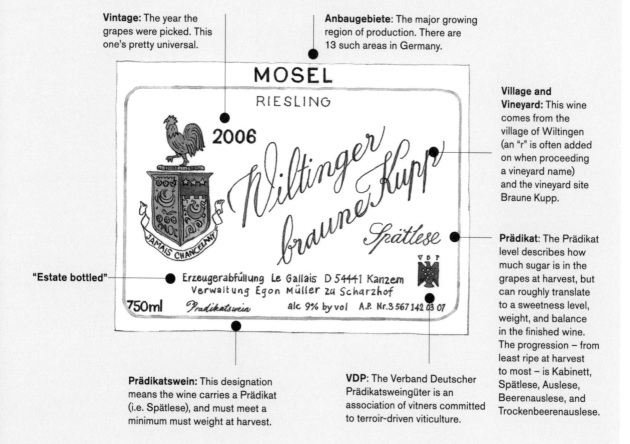

Village and Vineyard: This wine comes from the village of Wiltingen (an "r" is often added on when proceeding a vineyard name) and the vineyard site Braune Kupp.

"Estate bottled"

Prädikat: The Prädikat level describes how much sugar is in the grapes at harvest, but can roughly translate to a sweetness level, weight, and balance in the finished wine. The progression – from least ripe at harvest to most – is Kabinett, Spätlese, Auslese, Beerenauslese, and Trockenbeerenauslese.

Prädikatswein: This designation means the wine carries a Prädikat (i.e. Spätlese), and must meet a minimum must weight at harvest.

VDP: The Verband Deutscher Prädikatsweingüter is an association of vitners committed to terroir-driven viticulture.

122

Village and Vineyard: This wine comes from the village of Rüdesheim and the vineyard Berg Scholssberg.

Deutscher Qualitätswein: Formerly considered an inferior quality category compared with Prädikatswein – by German law, less ripe grapes are required and **chaptalization** is allowed. Now, all VDP dry wines are released as Qualitätswein, which has brought renewed energy to the category.

LEITZ
WEINGUT
2010
Rüdesheimer Berg Schlossberg
Riesling trocken
Deutscher Qualitätswein · Product of Germany
A. P. NR 24 079 034 11
Still Wine · Contains Sulfites

alc. 13,5% by vol · e 750 ml

"Winery"

Trocken: This wine is dry (as opposed to having perceptible sweetness). Legally, in Germany, there can be up to 9 grams per liter of **residual sugar** in a trocken wine, but acidity is so elevated that it still tastes quite dry.

AP Number: The Amtliche Prüfungsnummer number includes codes for anbaugebiete of production, village, producer submission number, and year of submission.

Alcohol: Looking at the alcohol level on a bottle of German Riesling is the most helpful and concrete way to determine if it is dry or sweet. Sweeter styles of Riesling will have alcohol levels below 9%. High-quality dry styles will often be at 12% or above. The increasingly popular category of feinherb wines can sit between these two points.

VIGNETTE

ABBAZIA DI NOVACELLA *SYLVANER* Alto Adige, Italy
BORGO DEL TIGLIO *FRIULANO* Collio, Friuli, Italy
PIEROPAN *LA ROCCA* Soave Classico, Veneto, Italy

Italian White Wine

Oh, the "somm trip." It all seems very glamorous.

You're flown out to a wine region, put up in (sometimes very nice) hotels, and toured around to the best vineyards and restaurants. The reality is not so glamorous, though I am probably more of a wimp about it than most.

The first night is always great. You check into your hotel, maybe take a quick nap, then everyone convenes for dinner and drinks. It's a new region to you, maybe even a new country. You're meeting new people, making new friends, and it's all a party. You finish the night with Negronis and stumble into bed around 1am.

Wheels up the next morning at 8:30am. You need to get up, get breakfast, and potentially get packed (depending on how mobile your itinerary is) by then. (Caffeine is a necessity for most on these trips, but I – alas – am intolerant of it.) You pull into the first winery at 9:30am. Right about now, you feel very lucky if you're in Alsace, Alto Adige, or the Mosel. You'll probably taste some white wine. But if you're in Bordeaux, the Douro, or Chianti Classico, tannic red wine will dominate the agenda, even at 9:30am. You'll spend a few hours at this stop, then hopefully it's lunch. There will be more wine. Two afternoon appointments are probably the minimum, then on to dinner.

Another 1am bedtime, and another 8:30am departure time. For multiple days. Throw in some region-wide tastings, where dozens if not hundreds of bottles are lined up, a few barrel tastings, which organizers erroneously assume is cool, and – if you're real lucky – a comparative tasting of different clones. By the end, your stomach will be raging with acid reflux, your teeth will be stained purple, and you'll be running on fumes. And when you get back to work – ragged, exhausted, bloated – everyone will ask how your vacation was.

But there are good parts too. The few days I spent in Alto Adige after my Germany road trip were invaluable. I got a sense of the culture of the region, tasted some phenomenal wine, ate the greatest food in the world (German meets Italian cuisine, yes please!), and met some people I still consider good friends. Other trips I've taken – to New Zealand, Margaret River, Chianti Classico, and Sonoma – have been some of the highlights of my career. But there are always moments when I wonder why I ever agree to these things. I look on with jealousy at fellow sommeliers with constitutions sturdier than mine, who thrive rather than decline on such trips.

Italian White Wine = The Answer for Difficult Pairings

In the wine world, there are what we refer to as "difficult pairings" – foods that give wine trouble. Whether it's a cut of red meat that's too tender for red wine, bitter vegetables that make wine taste astringent, or spicy foods that clash with alcohol, the answer lies in the white wines of Italy.

Bitter
Radicchio + Ribolla Gialla
Try: Doro Princic *Ribolla Gialla*, Collio, Friuli

Oily
Sardines + Etna Bianco
Try: Benanti *Pietra Marina*, Etna Bianco Superiore, Sicily

Fatty
Prosciutto + Friulano
Try: Borgo del Tiglio *Friulano*, Collio, Friuli

Smoky
Speck + Sylvaner
Try: Abbazia di Novacella *Sylvaner*, Alto Adige

Salty
Baccala + Soave
Try: Pieropan *La Rocca*, Soave Classico, Veneto

Tender
Veal shank + skin-macerated white
Try: Gravner *Bianco Breg*, Venezia Giulia, Friuli

Herbaceous
Pesto + Pigato
Try: Punta Crena *Vigneto Ca da Rena Pigato*, Riviera Ligure di Ponente

Cheesy
Gorgonzola + Marsala
Try: Marco de Bartoli *Vigna La Miccia Riserva*, Marsala Superiore Oro, Sicily

Fontina + Petite Arvine
Try: Les Crêtes *Petite Arvine*, Valle d'Aosta

Fried
Fritto misto + Fiano
Try: Ciro Picariello, Fiano di Avellino, Campania

Vinegary
Scapece + Franciacorta
Try: Bellavista *Vendemmia Satèn*, Franciacorta, Lombardy

Vegetal
Carciofi alla giudia + Verdicchio
Try: Sartarelli *Tralivio*, Verdicchio dei Castelli di Jesi Classico Superiore, Marche

White asparagus + Sauvignon
Try: Venica & Venica *Ronco delle Mele Sauvignon*, Collio, Friuli

Hot
Spicy vongole + Moscato d'Asti (really!)
Try: G.D. Vajra, Moscato d'Asti Piedmont

Napa Cabernet

I had expected to wake up the morning after the Advanced exam with a sense of relief and peace.

But an unsettled feeling had crept back in, one that I hadn't felt since before starting on Klonopin. Plath had her bell jar, but to me it felt like a layer of gauze lined the inside of my body. Sometimes this layer was thin, and just made me feel ever so slightly distant and disengaged. Other times, it was as thick as a quilt and I could barely see the words or faces in front of me. I felt small and foreign inside myself, and the world was a fuzzy, dark expanse. I never knew if I would wake up feeling like me, or like this tiny, removed, anxious, sad version of me.

I went to see a new psychiatrist in New York. She held the opposite views of my Chicago doctor, and was obviously concerned that I had been on Klonopin for so long. She told me that if people stay on Klonopin for long periods, their bodies become accustomed to it and it no longer works as well as it did. The anxiety comes back. At this point, I could either increase the dosage, thereby increasing dependency, or get off.

The way she described it, it seemed I only had one real option: to get off Klonopin. It seemed like the healthy thing to do for my body, and the idea of a clean slate was attractive. I thought I had conquered the adolescent fear and anxiety that had caused me to go on medication in the first place, and with the Klonopin gone, maybe I would go back to being "happy and carefree." (I convinced myself that there was a time when I was in fact happy and carefree.)

I embarked on the weaning process with my characteristic lack of trepidation, cutting my dosage in half on the first day. The next night, I woke up sweating with my heart racing. I was voraciously hungry and craved nothing but sugar. I couldn't see or walk straight and my skin tingled and burned. I wasn't scared, though. I knew exactly what was causing the symptoms. If anything, I was a little amused by how aggressive the withdrawal symptoms were. I went back up to three-quarters of a pill and evened out a bit. Within two weeks, I evened out a lot. I felt even better than I had at my best on Klonopin. *This withdrawal thing is going to be a breeze.*

During this time, I still thought about Jon, but I started seeing someone else. On our first date we bought oysters from Chelsea Market and drank Chablis on his rooftop. Megan called him Gingervitis (we were wont to come up with nicknames for the guys we dated). The nickname was a nod to his hair color, certainly, but she also joked that he might seem sweet at first, but would only cause inflammation.

—

I almost never get sick, but I had to leave work one Saturday with a 104° fever. Perfect timing, I thought. The online entrance exam to TopSomm (a national sommelier competition) was the next day, and I was supposed to fly to Napa the day after that for the Rudd Roundtable, a Masters prep weekend for the high scorers on that year's Advanced exams. Ginger brought me soup. I told him that I didn't want a boyfriend. He told me he respected this, but I could see he thought he would change my mind.

My fever broke early the next morning, and I enjoyed a few hours of blissful sleep. I took the online entrance exam for TopSomm (I had at least been too sick to get nervous), and the following day I flew out for the Rudd Roundtable. I was still exhausted and frail, but I wasn't going to miss this weekend.

Rudd is a winery based in the Oakville region of Napa. Owned by the late and great Leslie Rudd, the winery is dedicated to preserving the heritage of Napa and investing heavily in the education of the sommelier community. The weekend consisted of a blind-tasting of Oakville Cabernet, some social events with the Masters, and then workshops devoted to preparing for the theory, tasting, and practical portions of the Master Sommelier exam.

NAPA CABERNET

The Oakville tasting was illuminating. I had never drunk much in the way of Napa Cabernet, and I thought at the time that they all pretty much tasted the same. This tasting pitted some of the heavy-hitters of Napa against one another: Rudd (of course), Screaming Eagle, Harlan, Far Niente, Robert Mondavi, Heitz, and Opus One all make wines from this small slice of Napa. What the wines had in common was expense: none were cheap. For some, this expense could easily be attributed to high-quality fruit sources, showing up as complexity and depth in the wines. This expense could also be seen in the slick veneer of new oak. For some wines, the oak integrated magically, creating a broader and more regal whole. For others, the oak sat awkwardly on top, a halo of vanilla and coconut that would seem more at home in a tropical cocktail than a bottle of red wine. The expense of these wines could also be attributed to a myriad of factors including high-priced consultants, cutting-edge winemaking technology, soaring real-estate prices, and – perhaps most significant – *demand*.

I wondered at the mindset of wanting something simply because it is expensive and rare. It made me question my own *whys*. Did I want Jon because I couldn't have him? Did I *not* want Ginger because he had made it too easy for me? And why did I want to pass the Masters exam? Simply because it was a hard thing to do? And if I could answer all these questions in the affirmative, was this really such a bad thing?

2019 Classification of Napa Valley

The price a wine can command always creates an expectation of quality. At times in history, price and quality have been inextricably linked. For the 1855 Exposition Universelle of Paris, Napoleon III asked the merchants of Bordeaux to draw up a classification of the best Châteaux. This classification was based on the price of the wines as well as the reputation of the estates. Each Châteaux was ranked in one of five "growths," with first growths considered the premier producers.

Curious to see how this classification would extend to another region and hold up in modern day, I applied the same criteria to create a current classification for the Napa Valley. I've chosen 60 estates (the same number in the 1855 classification) as well as assigned the same number of estates to each "growth."

The criteria: The prices are based on the producers' flagship/premier wine. The wines must have some availability on the secondary market (i.e. available to buy if you're not on the mailing list), they must have been produced for at least ten vintages, and they must be at least 50% Cabernet.

Finally, just as one talks about the 1855 classification by dividing the Châteaux into their communes (Margaux, Paulliac, etc.), this classification is mainly split into **AVA**s. The exceptions are Pritchard Hill which, due to some copyright issues, is not technically its own AVA; "Napa Valley," which means the wine comes from a single tract of land, but it does not fall within a designated sub-region of Napa; and "Multi," which means that the grapes come from multiple AVAs throughout Napa Valley.

My takeaway with the Bordeaux and Napa classifications is that quality is incredibly high throughout, from first to fifth growths. The difference is more a stylistic one: the first growths are dense, concentrated, oaked, and extracted. As one travels down toward the fifth growths, the wines tend to have slightly less new oak, less purple fruit, and more rustic and savory notes can poke through in youth. The point is: find out what you like and drink that. More expensive doesn't always mean better, in Napa or Bordeaux.

2019 Classification of Napa Valley

Oakville
Screaming Eagle *Cabernet Sauvignon* **1st**
Harlan Estate **1st**
Bond *Vecina* **2nd**
Dalla Valle Vineyards *Maya* **2nd**
Futo *Cabernet Sauvignon* **2nd**
Opus One *Opus One* **2nd**
Paul Hobbs *Beckstoffer To Kalon* **2nd**
PlumpJack *Reserve Cabernet Sauvignon* **2nd**
Schrader *Beckstoffer To Kalon* **2nd**
Carter Cellars *Beckstoffer To Kalon "The OG"* **3rd**
Heitz Cellar *Martha's Vineyard* **3rd**
Lail Vineyards *J. Daniel Cuvée* **3rd**
Realm Cellars *Beckstoffer To Kalon* **3rd**
Rudd *Oakville Estate Red* **4th**
Robert Mondavi *To Kalon Reserve* **5th**

Spring Mountain District
Pride Mountain Vineyards *Reserve Cabernet Sauvignon* **2nd**
Philip Togni Vineyard *Cabernet Sauvignon* **5th**

Atlas Peak
Kongsgaard *Cabernet Sauvignon* **4th**

Diamond Mountain District
Diamond Creek *Gravelley Meadow* **4th**

Howell Mountain
Outpost *True Vineyard Cabernet Sauvignon* **3rd**
Dunn Vineyards *Howell Mountain Cabernet Sauvignon* **5th**

Mount Veeder
Lokoya *Mount Veeder Cabernet Sauvignon* **2nd**

Calistoga
Eisele Vineyard Estate *Eisele Vineyard* **2nd**
Larkmead Vineyards *Solari Reserve* **4th**
Chateau Montelena *The Montelena Estate* **5th**

Stags Leap District
Shafer Vineyards *Hillside Select* **3rd**
Stag's Leap Wine Cellars *Cask 23 Cabernet Sauvignon* **3rd**
Cliff Lede *Poetry* **4th**

Producers are grouped within their region by average price, with 1st growths being the most expensive (upwards of $750 per bottle) and 5th growths the least (around $150 per bottle – still quite expensive wines!). Don't be seduced or intimidated by Napa's high prices. Seek to explore the different terroirs and styles on offer to find what you like.

Pritchard Hill

Bryant Family Vineyard
   Cabernet Sauvignon   **1st**
Ovid *Red Wine*   **3rd**
Chappellet *Pritchard Hill
   Cabernet Sauvignon*   **4th**

Rutherford

Scarecrow *Cabernet
   Sauvignon*   **1st**
Staglin Family Vineyard
   Cabernet Sauvignon   **3rd**
Inglenook *Rubicon*   **5th**
Quintessa *Quintessa*   **5th**

Yountville

Kapcsándy Family Winery
   *State Lane Vineyard
   Grand Vin*   **2nd**
Dominus Estate *Dominus*   **3rd**
Blankiet Estate *Proprietary
   Red*   **3rd**

Napa Valley

Levy & McClellan *Cabernet
   Sauvignon*   **2nd**
Kenzo Estate *Ai*   **3rd**

St. Helena

Abreu *Madrona Ranch
   Cabernet Sauvignon*   **2nd**
Colgin *Tychson Hill
   Vineyard*   **2nd**
Hundred Acre *Kayli Morgan
   Vineyard*   **2nd**
Grace Family Vineyards
   Cabernet Sauvignon   **3rd**
Vineyard 29 *Estate Cabernet
   Sauvignon*   **3rd**
Melka Estates *Métisse
   Jumping Goat Vineyard*   **4th**
Corison *Kronos Vineyard*   **4th**
Hourglass *Cabernet
   Sauvignon*   **4th**
Spottswoode Estate *Cabernet
   Sauvignon*   **5th**
Tuck Beckstoffer Wines
   Dancing Hares   **5th**
Barbour Vineyards *Cabernet
   Sauvignon*   **5th**
Revana Family Vineyard
   Cabernet Sauvignon   **5th**

Multi

Joseph Phelps Vineyards
   Insignia   **4th**
Caymus Vineyards *Special
   Selection*   **5th**
Corra *Cabernet Sauvignon*   **5th**
Far Niente *Cabernet
   Sauvignon*   **5th**
Favia *Cabernet Sauvignon*   **5th**
La Sirena *Cabernet
   Sauvignon*   **5th**
Lewis Cellars *Reserve
   Cabernet Sauvignon*   **5th**
Pahlmeyer *Proprietary
   Red*   **5th**

BARBOURSVILLE VINEYARDS *RESERVE NEBBIOLO* Monticello, Virginia, USA
CADUCEUS CELLARS *CHUPACABRA* Cochise County, Arizona, USA
L. MAWBY *BLANC DE BLANCS* Leelanau Peninsula, Michigan, USA

The Other 46 States

It was Super Bowl Sunday 2014, and I was in the mood to do something reckless.

I only remember the date because I was at a friend's house "watching football" (but really just scrolling through my phone), when a message popped up from him. Not Jon or Ginger, a different *him*. He was a Master Sommelier who lived across the country, and within a few days of us messaging, he had invited me to fly out and stay with him for a weekend. I'd never met him. We became acquainted in the most modern of romances: through Twitter. A comment here, a retweet there, and finally a direct message. We carried out a flirtatious social media and text relationship, which led to an invitation to "show me around his city." In one of my ballsier *fuck-it-why-not* moments, I agreed to fly out. Something fun and frivolous to wash any thoughts of Jon off me – and I imagined we'd drink some really good wine.

My friend Tom told me to be cautious. "You don't want anyone to think your accomplishments are due to who you've fucked rather than what you've done."

VIGNETTE

My sister thought I was crazy, as usual. "You've never met this guy before?!" Beth huffed. She made me send her his name and address and told me if I didn't check in every few hours, she would call the local police. She was less concerned about my reputation being tarnished, and more concerned about my tarnished teeth being found in a dumpster six months later.

Despite all protestations – and all reason – I bought the plane tickets. They were for late February.

Coincidentally, that same weekend (hospitality weekend – a Monday) the Masters theory exam was being offered in Houston, San Francisco, and Atlanta. I was a year away from taking my theory exam, but many people I knew would be sitting for the exam in Houston. Tom. Jon. Even Ginger. All in one room together. I thought of Jon's exam the previous year, how I had cheered him on through it all, and how he had come back and taken solace in my arms.

I harbored a secret hope that Jon would somehow find out about my weekend with Master Stud (MS), as Megan called him, and be insanely jealous. I also hoped, in my ugliest of moments, that Jon would fail the exam.

MS told me how excited he was for my visit. He told me about the plans he had made for us to go on hikes and drink spiked cocoa by a fire. It was Friday and I was flying out Monday. In another supreme coincidence, my flight out connected through Houston. I imagined a series of events in which I got stuck in Houston due to weather and ended up at Jon's hotel, once again consoling him over an exam failure. I pushed these thoughts away, and steeled myself for what would hopefully be a distracting respite of a weekend.

The next day, Saturday, MS texted me with "an idea."

"How about you stay in Houston instead? I'll meet you there, and we can party with the other MSes."

Party with the other MSes? It was one thing to have a perhaps not-so-secret rendezvous that was completely private and personal, and quite another to show up with him to a Court exam – one that I would be taking in a year's time. How could I party with the Master Sommeliers who would judge my exam the following year? And on top of it all, both the man I was seeing and the man I was probably still in love with would be there. Though MS didn't know this last bit, the preceding was enough reason to keep me out of there. *Was he nuts?*

I immediately sent a screenshot to Tom and could see him laughing on the other end. Tom confirmed that *yes, that is the craziest thing ever* and there was no way I should agree to it. But

133

what to say to MS? *Are you out of your fucking mind?* seemed ill-advised.

"Well, yeah, I mean, that does sound like fun," I wrote back. "But I know a number of people taking it, and I feel like it's disrespectful to show up to party when many of them probably won't pass."

"Everyone needs a drink at the end of the day!" he retorted.

"Yeah, sure, but I also feel like it's weird for me to show up to a test that I'm going to take next year," I came back. "The MSes can't party with me this year and judge me next year. I feel like it's unprofessional."

"We MSes are just people, like anyone else!"

This last text felt like a quote from *Zoolander*, as Derek tried to explain to Matilda that he was just like her (except, you know, a lot, lot prettier).

I couldn't help feeling that this was his version of cold feet. Something about our recent interactions had convinced him that he didn't want to spend a weekend alone with me. And he was either dumb and thought that being around a group of people in our field would minimize any awkwardness; or he was real smart (more likely) and knew I would never go for it, and this was his way out.

Well, I bit. I told him that if he wanted to go to Houston so bad, he absolutely should, and I would cancel my plane tickets. He feigned disappointment at not seeing me and left our conversation with "Another time!" – like he'd cancelled a brunch date. We both knew there would not be another time. I was confused and disappointed and took it way too personally. I already had the time off from work, though, and Megan convinced me to spend a few days alone at her mom's house in Connecticut: take baths, study, do yoga, stick pins in voodoo dolls, etc.

That Monday was the theory exam in Houston. Tom didn't pass. Neither did Ginger. Nor Jon. I felt for them, and I feared for myself and how I would fare next year. I talked on FaceTime with Ginger and, like a spool unthreading beyond my control, the story of the last few days fell out of my mouth. He had always been there for me in tough times, and a really idiotic part of me was hoping he could be the shoulder I cried on after being slighted by MS. I soon realized how misguided and hurtful this was. He pursed his lips and nodded, fully understanding what I had so casually blurted out. He knew then that I really didn't want him to be my boyfriend. And that I had used him. His eyes glassed up and, just like that, we were done.

I fell in and out of sleep that night. I couldn't tell if I was dreaming or awake, but I kept staring at the closet door of the room I was staying in. I was convinced that the bathroom on the other side of the wall occupied the same space. I got out of bed (or maybe I dreamt that I did), left the bedroom and went into the bathroom. It was directly on the other side of the wall and extended its length. I came back into the bedroom and stared at the door of the closet. If it opened, it would be occupying the exact same space as the bathroom. There was no way around it. They couldn't both exist.

I opened the door to the closet and found a dark and cavernous room. The walls of the closet were lined with bottles of wine, with labels and names that were unfamiliar to me. I picked up the first bottle. Texas wine. The second. Colorado wine. The third. Virginia wine. The fourth. Arizona wine. I slammed the door and crawled back into bed, turning my back on the demons of the night.

I woke up the next day with my heart racing, dizziness washing over me even as I lay in bed. My chest stung with a heat that spread outwards, coursing like venom through my blood. I was exhausted, but so wired that I couldn't go back to sleep.

I looked towards last night's closet and found a blank wall with a vaguely colonial piece of art on it. No door. No closet. I spun around to the other side of the room, where the closet actually was, no bathroom impeding its spatial veracity. I jumped up and opened the door of the closet. No shelves of wine. I shrugged it off – *what a weird dream* – and just as I was about to close the closet door, I saw a gift bag on the floor containing two bottles of wine. The tag, seemingly from a proud neighbor, read: "CT makes good wine too!" And inside the bag, a bottle of blueberry wine and a bottle of Merlot, both from Connecticut.

THE OTHER 46 STATES

Why Not the Other 46 States?

Every American wine region has its viticultural issues: Napa has had catastrophic floods. Washington State endures water shortages. The Willamette Valley suffers from rot. New York's Finger Lakes can freeze over. But in general, for these regions (especially the first three), the conditions are quite favorable for wine production. The benefits outweigh the hardships. And prices can be fetched that allow vintners the freedom to deal with issues as they arise.

For the other 46 states, the scale tips in the opposite direction. The conditions working against them are more laborious and costly than in places like Napa and Sonoma, and the favorable conditions do not generally create wines that rise to the same strata (and fetch the same prices).

	OHIO	MICHIGAN
Issues	The most active **AVA** within Ohio is the Lake Erie AVA. While the lake moderates the cold ingress of the winter, extending the growing season and warding off frost, the moisture it contributes in the warm months makes for some of the most intense mildew and rot pressure in the country.	Michigan's proximity to lakes helps moderate what would otherwise be a winter that no vine could survive. In the warmer months, the moisture from the lake results in mildew and disease pressure. Yields can be drastically reduced in years where growers see frosts on both ends and significant rot in the summer.
AVAs of Note	Lake Erie	Leelanau Peninsula, Old Mission Peninsula
Most Succesful Grapes and Styles	Pinot Noir, Grüner Veltliner, Chardonnay	Ice wine from Riesling and Vidal, sparkling wine from Chardonnay and Pinot Noir, Gewürztraminer, and Cabernet Franc

But a revolution is underway. As viticultural know-how and technological prowess grow, vintners are dealing more effectively with issues like frost, disease pressure, and short growing seasons. The character created by the complex geological compositions and unique geographic features of these regions is starting to emerge from behind the cloud of their handicaps.

TEXAS	COLORADO	ARIZONA
Texas often experiences summer hail storms that tear fruit to the ground and bruise vines. These wounds can allow disease to step in. Late spring frosts can also be an issue. Furthermore, Texas is prone to summer deluges that flood vineyards.	The issue in Colorado is the shortened growing season. Colorado is often replanting vines that die from freezing temperatures, and dealing with late spring frosts that drastically reduce yields and hinder fruit formation altogether.	Arizona is quite dry and access to water is a struggle, both in the vineyard and in the winery for cleaning. For wineries built above ground (most of them), evaporation is an issue. Producers have moved away from smaller barrels as they don't hold temperatures well and the evaporation rate is quite rapid. The hot, dry climate is literally making their wine disappear!
Texas Hill Country (includes Fredericksburg and Bell Mountain AVAs), Texas High Plains	Grand Valley, West Elks	Sonoita, Willcox, Verde Valley (the latter is a notable region but not an AVA)
Tempranillo, Sangiovese, Viognier	Merlot, Cabernet Sauvignon, and Syrah in Grand Valley; Riesling, Gewürztraminer, and Pinot Noir in the cooler West Elks	Cabernet Sauvignon, Merlot, and Rhône varietals

138

Red Burgundy

Until this point, I remained convinced that my withdrawal from Klonopin would be easy.

But I woke up that Tuesday morning in Connecticut to what would become a new era of my life.

I spent the day studying, reading and rereading my book on Burgundy. My eyes hazed over and I blinked violently to restore them, but focus eluded me. The intricacies and nuances of Burgundy lodged themselves in the painful, exposed parts of my brain. The transience of site in Burgundy – how everything can change with the depression of a slope, the shift of exposure, the transition of soil – made me hopeful that tomorrow could be different.

After a day of fighting through the heavy fog that hung over me, my body finally gave in to a nap on the couch. I found strained and tenuous relief. I went to bed that night assuming the day was an anomaly, so naïve and unaware of the struggles to come.

The next day opened much like the one before, with a suffering that would become all too typical. I took the train to New Haven to see a few friends who went to grad school at Yale. The night started early, with Manhattans at 5pm at my friend's apartment. Dinner brought more cocktails and wine. I ordered a bottle of expensive red Burgundy for the table, wanting to see if I could impress on myself the optimism of yesterday's study. But the venom that had heralded the day just burned more strongly.

I was planning to stay the night and take the train back to Manhattan the next day, but the thought of sleeping restlessly on someone else's couch was suddenly unbearable to me. On the train that night, the voltage became even more amplified, on the verge of combustion. I texted Jon.

"Sorry to hear you didn't pass."

"Thanks."

"Are you back from Texas?"

"Yeah."

I proposed we meet up. He agreed. We met at Milk & Honey, the same bar where he told me he could one day fall in love with me. We had a drink. I think we also might have had shots, at my request. I remember very little of the night, my withdrawal symptoms meeting the copious amount of booze at a dangerous point. He reiterated that he wasn't looking for a relationship. I brought him home anyway.

In the weeks that followed, I went through two sets of withdrawals. The one from Klonopin continued to tear a monstrous hole through my life. The withdrawal from Jon was just as devastating. At least I understood the carrots that Klonopin had offered me – it had soothed my anxiety, made everything a little softer and less angular, and helped me focus. The good times with Jon were so few and so removed that I couldn't understand my addiction. My body and emotions craved him in a way that my mind couldn't comprehend.

I went to tasting groups he hosted. I dropped by the bar at Eleven Madison Park with friends. I found reasons to put myself in the same room as him. I wasn't ready to give him up completely, and I wondered if I ever would be. I kept sifting through my mind and body to find those subtle transitions of site that led me to this, and those that would lead me out of it.

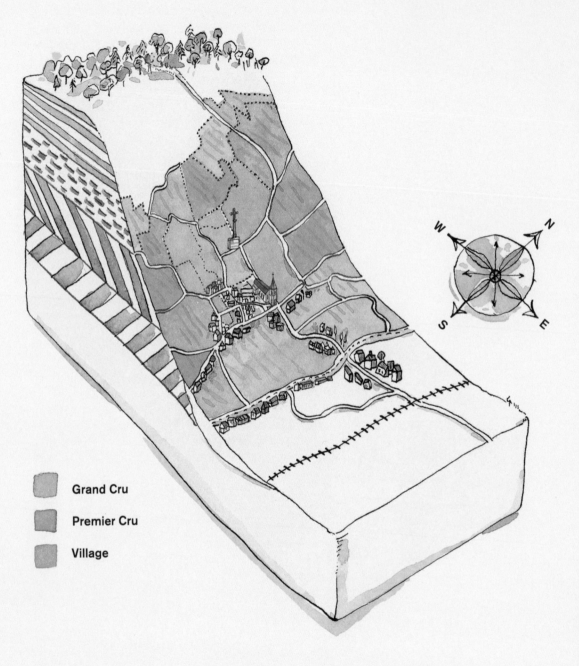

Grand Cru

Premier Cru

Village

VIGNETTE

Anatomy of a Vineyard Site

The vineyard land that makes up Burgundy is roughly split into four tiers (with some exceptions and gray areas, as always). The best sites, making up about 2% of production, are **Grand Cru** vineyards. **Premier Cru** comes next, comprising approximately 12% of the production. **Village** is the third tier (pronounced VILL-ahj), which describes wines coming from one commune (i.e. Vosne-Romanée, Gevrey-Chambertin, or Meursault). The fourth tier are Regional wines, which can fall under a few different appellations, Bourgogne being the most prominent.

Aspect	In order to maximize sun exposure, most top-quality European vineyards face south (at least partially). The Côte d'Or – the main slope of Burgundy – faces southeast, with some variation among individual vineyard sites.
Soil Composition	The general soil types of Burgundy are a mixture of clay and limestone, often referred to as marl. The composition of top soil and bedrock, as well as depth of top soil, can make a large difference in the expression of the wine.
Slope	Too high on the slope and the top soils are quite thin, creating wines that are a bit lean and austere. Too low on the slope and the soils are dense and marshy, creating wines with less focus and precision. Mid-slope is optimal.
Row Orientation	Most vines in Burgundy are planted east-west for even sun exposure and air-flow purposes.
Yields	In general, the less fruit produced, the higher quality that fruit will be. Think of it like this: the vine has finite resources, and the fewer grapes, the more concentration each one has. Yields usually correlate inversely with planting density: the more densely planted a vineyard site, the lower yields it will realize.
Age of Vines	As vines mature, they produce more concentrated fruit, but less of it. They are also more resilient to disease and weather. Many argue the older the vines the better; others contend that there is a limit to vine age (both in terms of financial viability and ideal expression).

ETIENNE DUPONT *CIDRE BOUCHÉ BRUT DE NORMANDIE* Normandy, France
ROYAL TOKAJI COMPANY *MÉZES MÁLY* Tokaji-Aszú, Hungary
VINHOS BARBEITO *FRASQUEIRA BUAL* Madeira, Portugal

Cheese Wine

Competitions have always been an
important part of my preparation.

The first sommelier competition I participated in was in 2012 while I
was living in Nashville. It was Top New Somm, put on by the Guild of
Sommeliers, and I was competing in the Southern regionals. I had not
yet taken my Certified exam, I had never done a blind flight of six wines,
and my theory knowledge was still patchwork. I surely got the lowest
score of the day, but I now knew where the bar was. I also met some
amazing sommeliers who would become my friends and colleagues.

In 2013, I made it to the national finals, winning the Southern
regional competition. Once again, I almost certainly got the lowest
score in this elevated bracket. But it was progress. In 2014, I made
it to the New York regionals, but not to the finals.

That same year, I competed in the Chaîne des Rôtisseurs Jeune
Sommelier competition. I beat two other New York sommeliers to
make it to the nationals in Carmel, California. I was excited to go out
there and compete, but my body did not want to cooperate. I woke
up early in the morning the day of the competition to a panic attack.
My heart was racing, my blood burning, my body throbbing as I lay
silently in bed. I took half a Klonopin to go back to sleep. I got myself
to the competition on time, groggy, with a pounding subterranean
anxiety. I smiled and hugged my old friends. When they asked me
how I was, I responded with a smile.

"Everything's great. You know, same old."

I focused hard and pulled myself through the different activities asked of me throughout the day: a written theory exam, an oral tasting, wine-list correction, blind spirits identification, a pronunciation exercise, and three separate service tables where I opened Champagne, decanted, used a **Coravin**, answered questions on beer and sake, made cocktails, and paired with a tasting menu (cheese pairings, always cheese pairings at competitions). These were things I enjoyed, and things I was good at, but my spiking anxiety clouded my pleasure and precision. By the last service table, my adrenaline was so rampant I had trouble focusing at all. I presented the ordered bottle of wine to the wrong person – twice – and forgot information that should have been on the tip of my tongue.

The proctors at the table looked down and rapidly ticked off points, surely judging me as sloppy and unfocused.

Much to my surprise, I won the competition. Hundreds of members of the Chaîne des Rôtisseurs were in attendance that night, and everyone wanted to congratulate me. It was my night, and I felt a slurry of emotions. I was certainly happy and proud of myself. But I didn't understand why it had to be so hard, and why I felt so defeated by the day.

Before I left that night, one of the judges on my final service table, a Master Sommelier, came up to me. "I know your service is better than what I saw today. But, hey, I guess it was good enough."

—

The international finals of the Chaîne des Rôtisseurs competition were held a few months later in Denmark. I got into Copenhagen a day before the competition started. Between the jetlag and the anxiety of competing, I needed extra Klonopin to get me through the night. I woke up groggy, but at least not overcome with anxiety. The initial rounds of this competition were much easier than what we had experienced in the US. The national competition in Carmel had been brutal – the highest theory score was just over 50%. This test was much more reasonable and helped me build some confidence going into service.

Service was straightforward too, yet my hand still shook during both my Champagne and decanting tables. The judges gave me surprised looks: I knew the material and I came off confident – *but why is she shaking?*

On the second day of the competition, the judges announced the top three contestants. I was one of them. I was pleased to be in this group, but not pleased about competing in the final round, which would take place in front of an audience. I was the second to compete, and I waited outside the door while the first contestant went through. I had read enough about shaking at this point to know that trying to fight it only makes it worse. I stood in the hallway and shook my arms and legs out, lightly at first. I repeated over and over in my head *let them see you shake.* I started shaking out my limbs more and more violently. I shook and shook and shook until my body vibrated and my head ached. *Let them see you shake let them see you shake let them see you shake...*

I was called in after 15 minutes or so. Much to my surprise, and relief, I didn't shake in front of the audience. I made a cocktail; I decanted a bottle of wine; I answered rapid-fire theory questions, poured Champagne, and blind-tasted. I took a huge shot of Fernet when I was finished and vowed to enjoy myself for the rest of the trip. I was awarded second place that night, and I couldn't have been happier.

After the competition in Copenhagen, I stayed in the city for a few days. I ate and drank well. I met a Danish man with a thick beard and a gentle smile who showed me around. It was a magical few days, a reprieve from thinking about my health, and the chance to enjoy my life. It was something I hadn't done in a very long time.

On my last day in Copenhagen, the symptoms came back. It was like my body knew that this vacation was over, and it was time to go back to the struggle.

A New Way to Look at Cheese Classification and Pairing

To talk about cheese, I've enlisted my friends Bronwen and Francis Percival, veterans of the wine and cheese world, curious and generous academics, and authors of *Reinventing the Wheel: Milk, Microbes, and the Fight for Real Cheese*. I met Francis at a wine conference in 2016, where he presented one of the most compelling takes on cheese I'd ever heard.

Cheese (and its respective wine pairings) can seem like a daunting amount of information to learn. There are hundreds of types of cheese to know…and thus hundreds of individual pairings to master. The Percivals simplify this. They argue that it is possible to understand all cheeses through the prism of just two factors: their level of moisture and their acidity. Cheese, as we will see on the chart on the next page, can be classified neatly in one of four quadrants – with one additional for pairing purposes.

A New Way to Look at Cheese Classification and Pairing

High

ACIDITY

Low

Cheddar and Friends

With the cheeses in the other quadrants, acidification and moisture drainage take place consecutively; with Cheddar and the like, these processes occur at the same time. This fact makes these styles the most labor- and time-intensive in cheesemaking.
Famous Examples: Cheddar, Cheshire, Lancashire
Best Pairings: Madeira, cider, Beaujolais

Alpine Cheeses

Alpine cheeses, or "herder cheeses," require vigorous effort over a very short period of time. They were often made by herders or other nomads, people with the opposite lifestyle and demands of the farmer's wife. These cheeses develop a malleable texture that can dry and crystallize with age, and flavors often described as nutty and sweet.
Famous Examples: Gouda, Gruyère, Comté, Parmigiano-Reggiano
Best Pairings: Vin Jaune, Amontillado and Palo Cortado sherries, white Bordeaux

Blue Cheeses

For pairing purposes, we've added a fifth quadrant (quintant?) to encompass blue cheeses. Blue cheese is defined by the addition of the mold *Penicillium roqueforti*, which becomes an important flavor component. These cheeses can range in terms of acidity and moisture levels, but always have a characteristic saltiness and sharpness.
Famous Examples: Roquefort, Blue Stilton, Gorgonzola
Best Pairings: Port, Sauternes, **Tokaji**, peated Scotch

Lactic Cheeses

Lactic cheeses – "farmer's wife cheeses" – are acidic, dense, and flaky. They take longer to make, as the acidity takes time to develop, but don't require a lot of effort throughout the process.
Famous Examples: Chevre, Valençay, Époisses
Best Pairings: Loire Valley Sauvignon Blanc, dry and off-dry Riesling, rosé

Creamy Cheeses

The low-acid/high-moisture cheeses are the easiest and quickest to make: no acidification needs to occur, and very little moisture needs to be removed. Without the acid promoting calcium dissolution, these cheeses also have the creamiest and smoothest texture.
Famous Examples: Mont d'Or, Brie, Harbison
Best Pairings: California Chardonnay, white Rhône wines, Champagne

Low   **MOISTURE**   High

Post– Competition Drinks

147

The aftermath of competitions is always an interesting scene.

Everyone's ready to let loose, and no one's too upset about the outcome of the day (unlike at exams). The drink of choice is typically a Negroni. Equal parts gin, sweet vermouth, and Campari, the Negroni is bitter and sweet, easy and challenging at the same time. It is a cocktail that masquerades as a light, aperitif-style drink, but it is actually quite boozy. Rounds of Negronis have been ordered after nearly every competition I've taken part in – to celebrate, to unwind, to forget.

My most memorable post-competition experience that year, though, was not for my own competition. The same weekend that the Chaîne des Rôtisseurs competition was held in Carmel, Jon

POST-COMPETITION DRINKS

was competing in the nationals of TopSomm in San Francisco. My competition was on a Saturday. I was attending a wedding Sunday evening in Berkeley, and then had a flight out of San Francisco to New York at 6am on Monday morning. I decided it made the most sense to drive into San Francisco Sunday night after the wedding so I didn't have to commute for as long the next morning. Plus, I had a place to stay. A friend of mine, Morgan, was also competing in TopSomm and said I could crash in his hotel room for a few hours. I told myself this had nothing to do with seeing Jon. It was just the smartest plan (obviously).

I got into San Francisco in time for most of the TopSomm afterparty. There were lots of good bottles being passed around and, of course, a bottle of Campari loitering in the ice-bath. I was immediately glad I came, and it had nothing to do with Jon. In the room were some of the greatest sommeliers in the country. People who had inspired me, mentored me, collaborated with me, and been my friends. Jon and I chatted briefly and curtly. As I made my way through the party, he haunted my periphery, but I tried to focus on having fun and socializing with others.

Jon said goodnight before he went to bed. I nodded back. My mind was too thick with regret and confusion to say much of anything. It seemed like Jon had something to say too, but we walked away in silence. Morgan also retired, telling me to knock on his door when I was ready to get some sleep. I stayed up with a few friends I rarely got to see, but within an hour headed down to Morgan's room.

I knocked and waited. No answer, not even a stir. I knocked again, harder this time. Nothing. I pounded on the door. No sound. I called his cell phone and it rang and rang.

I figured I had a few options. I could go back and hang out with my friends, and probably drive to the airport without sleeping at all. Or I could call Jon. I made what I told myself was the logical choice.

"Hello?" a sleepy voice said on the other end of the phone.

"Hey. Morgan's not answering his door. Or picking up his phone. Can I crash with you for a few hours?"

"Yeah, of course."

Jon gave me his room number. I knocked softy on the door and he opened it immediately, the light of the hallway causing him to squint. He always looked cute when he'd just woken up. Sort of confused, but harmless and soft. He stood there in his boxers and t-shirt and ushered me in, immediately climbing back into bed. I took off my pants and put on a pair of shorts. I got in the other side of the bed and turned my back to him.

As much as I wanted to be around Jon, I did not want to sleep with him again. Something about him not deserving it. And it opening old wounds.

He wrapped his arms around me and held me. He held me closer than ever, and as we lay there, he squeezed me harder and harder. He wrapped his top leg around my body and tightened his grip. I clung to him with equal enthusiasm, never wanting to fall asleep. It was the most euphoric feeling, being in his arms, and I was afraid I would never feel it again. He said something as he was falling asleep, about wishing there was a way we could always be in each other's lives. I wished that, too. And didn't understand why we couldn't.

149

POST-COMPETITION DRINKS

Wheel of Negroni Variations

The balance of a Negroni depends on having one spirit (gin), one aromatized wine (vermouth), and one bittering agent (Campari), but the classic components can be swapped out for others. The below wheel is just the beginning of the possibilities for Negroni variations. Take a spin.

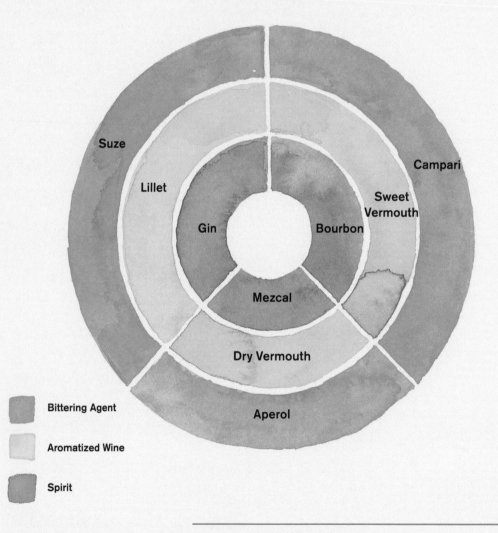

Suze

Lillet

Campari

Sweet Vermouth

Gin

Bourbon

Mezcal

Dry Vermouth

Aperol

Bittering Agent

Aromatized Wine

Spirit

Chablis

Why – I was asked (by others and myself) – would someone with a severe anxiety disorder who is undergoing medical withdrawal agree to be on television?

I don't have a good answer for this. I thought maybe the overexposure to anxiety-inducing situations could help me get over it. I thought maybe, in the long run, it could be good for my career. Or perhaps I just have a hard time saying no.

Whatever it was, I found myself at another competition in 2014. This time, it was organized for the pilot of *Uncorked*, a six-episode television show about candidates preparing for the Master Sommelier exam. It was a service competition at Eleven Madison Park just for the participants in the show: me and five other New York-based sommeliers. All men. Mostly my senior. All of whom had passed at least one section of the Master Sommelier exam.

I was taking a quarter pill of Klonopin each day now. I had never reached a stasis going from three-quarters to a half, but I figured I just needed to keep moving through it. And I thought things would be better when I got off of it entirely. *This has to be as bad as it gets.*

The gauze was thick and scratchy that day, rubbing me raw inside and out. The maître d', as we like to call the MC in service competitions and exams, pointed out how shaky I was going into my scene. I made a joke of it and walked into the frame, just wanting it all to be over.

The first exercise of the competition was a blind-tasting for quality. I was greeted by four Master Sommeliers and three glasses of wine. They told me all three contained the same grape from the same general region and asked me to identify it. I picked up the first glass by the stem, steadying the base with my other hand to keep it from shaking. I put my nose in and took a sip. Pretty neutral nose, **leesy**, long, high acid. The other two confirmed my initial impression: these were Chardonnays from Chablis.

"Very good," Geoff Kruth said. "Now one is a **Grand Cru**, one is **Premier Cru**, and one is **Village**. Please identify which is which."

I went back to the wines. I was looking for increased lees expression, more concentration, more alcohol, and more oak toward the Grand Cru side, with an overall lighter, leaner, softer feel on the Village end of the spectrum. I did my best to make distinctions, but my body was pulsing with distracting thoughts and energy. I eventually made my calls, which were apparently wrong when I looked at the judges' faces. Geoff reached for one of the glasses. Then another. Then the third.

"I think we poured them wrong," he said. Laura Maniec sat forward and began picking up the glasses as well.

"Yes…maybe," she chimed in.

They hummed and hawed for a moment, revisiting each glass. They agreed, they had been poured wrong.

They re-poured and returned the glasses to the table. It was now apparent to me which wines *should have been* in each spot. I revisited the wines and thought I could detect the differences, but the fog caused by my anxiety, the judges' gaze, and the watchful cameras was overwhelming my senses. I called what I thought *should* be right. The judges' faces were much more favorable this time.

I took a deep breath and moved on to the service portion of the competition. They began asking about Champagne. I surprised myself with my tableside demeanor; my nervous energy had transformed into friendly exuberance. It was almost as if my capacity for compassion had widened, and I could feel the judges responding to my very *humanness*. I was as vulnerable as I'd ever been that day – in front of four Master Sommeliers and even more cameras – and for a brief moment, I had fun.

That moment was quickly squashed as I reached the last segment of the competition: decanting. I had trouble getting the capsule off the bottle and could feel the shakiness rise in me. I could also feel the heaviness in the air as everyone wondered what was taking me so long. When I finally got the bottle open, I rested its lip against the glass of the decanter while I poured in the wine. This is a big no-no in decanting (the wine bottle shouldn't touch the decanter), but I preferred it to my hand shaking and spilling wine all over the **gueridon**. I choked up on the neck of the decanter as I poured it in the glasses around the table, trying to give myself as much control over the glass vessel as possible. I still shook as I poured for the judges – what I would describe as violent shaking but was later described to me as "barely noticeable." I thanked the judges and left the room as quickly as possible, hardly making it outside before I started to cry.

I craved some understanding. At times, I wished I could just wear a name tag – *Hi, my name is Jane and I am going through Klonopin withdrawal.* The burden of dealing with the withdrawal was compounded by the burden of having to pretend that everything was okay. I didn't want to hide it. But I also didn't think it was appropriate to announce it to the world. *Would anyone understand? Would anyone care? And would it help the ultimate goal – of me feeling better?*

I pulled myself together to go back for the results. I sat with the other five contestants in the lounge at Eleven Madison Park while the judges talked at a table no more than 50 feet away. We could hear some of what they said, but mainly just words here and there. My name came up and I cringed, expecting them to question why I was even on this TV show. How I could be a candidate for the Master Sommelier exam.

"She was great."

"She'll definitely be an MS someday."

"She had good energy."

I was shocked. Not only had I done well, it turned out that I had won. I couldn't figure out what lesson to take from the day. It was incredibly rewarding to know that all my hard work was paying off. But I was absolutely destroyed. My body was wrecked. The expended adrenaline circled through my blood like poison, draining my skin of color and my eyes of life. I spent the rest of the day practically comatose in my apartment. I tried to study but could not even sit upright in my chair. I tried to sleep but my heartbeat rung through my body. I settled for watching a movie on the couch, allowing myself a night off from studying and a reprieve from the pressure to feel good.

Les Clos: Then, Now, and Beyond

When starting in the wine industry, I assumed every important wine had been discovered and defined. That every classic wine region had long realized its majestic birthright. When I expressed this to a mentor and friend, he politely told me, "I totally disagree and, honestly, the opposite is almost certainly true." He gave me many antitheticals, including that in the 1950s people skied down Les Clos, a Grand Cru in Chablis. This one stuck with me, and I like to think about it when I feel mired in perceived permanence.

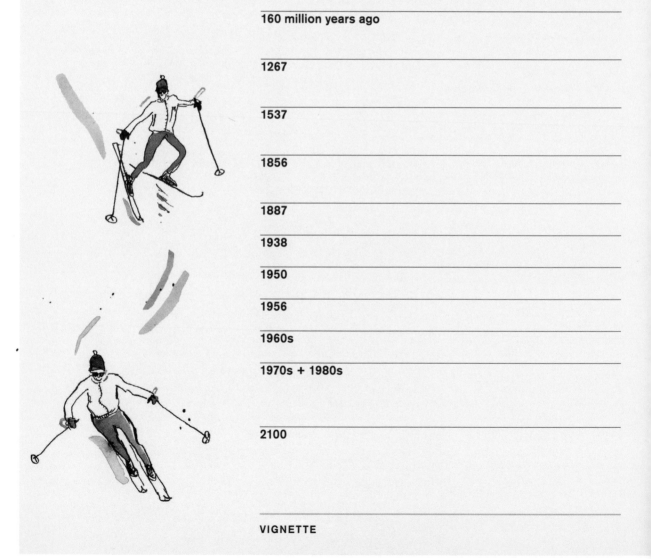

160 million years ago

1267

1537

1856

1887

1938

1950

1956

1960s

1970s + 1980s

2100

VIGNETTE

155

The sea withdraws in the Upper Jurassic Age, depositing the Kimmeridgian subsoil that makes the region so famous.

The first record of vine plantings in the region, though it is thought that vines could have been planted as far back as the ninth or even third centuries.

Les Clos receives its name, alternately spelled "Le Clou," "Les Clous," and "Le Clox." "Les Clos" refers to an enclosure in French. It is believed that the vineyard was enclosed by a wall at this time.

Railways open across France. The advantage Chablis formerly enjoyed in shipping its wines to Paris becomes negligible with the new accessibility of transportation.

Les Clos is devastated by **phylloxera** and **oidium,** which wipe out the vineyard completely.

The Chablis Grand Cru appellation is ratified.

There are only 500 hectares of vines in all of Chablis, compared with 40,000 hectares in the 19th century.

An incredibly brutal and cold winter; all vines die, and locals ski on the Grand Cru slope.

The revolution begins, with high-quality plantings and more advanced vinification practices.

René Dauvissat, François Raveneau, and Christian Moreau begin exporting their wines to the US and across the world, spreading the name of Les Clos near and far. Les Clos and the other **climats** of the Grand Cru slope reclaim their rarified place in the wine world.

Who knows what's in store for this vineyard? Will it be swallowed up by the ocean again as the currents change due to global warming? Will the ozone layer thin out, making the region more suitable for Marsanne and Viognier than Chardonnay? Or will the vines continue to tick along, making some of the most precise and filigreed Chardonnays in the world?

CHABLIS

156

Calvados

The Masters Exam took place in Aspen in May of 2014.

If Jon had passed theory back in February, he would be there. It was where I hoped I would be in a year. I took it as an excuse to contact Jon. *Have you heard from anyone? Did Morgan pass? Dana? Hristo? Pascaline?*

I had been out to dinner with Tom that night, eating at a midtown steakhouse hoping to court me as its wine director. Tom and Jon knew each other through tasting groups and industry events. Tom thought Jon was a good guy and could be a good guy for me; he just needed to come to his senses. "What's wrong with that boy?" Tom would say in his endearing Harlem-meets-Nashville way. He suggested the two of us meet up with Jon for a drink that night. We went to Milk & Honey.

Tom ducked out early, which isn't normally his style. Jon and I finished our drinks, cuddled in the same side of a four-person banquette table. "I guess we should go our separate ways," I said, unconvincingly.

"I don't really want that," he replied.

—

"I think he's changed," I said to Megan, when I agreed to go out on a date with Jon two weeks later. She gave me a look that kindly and empathetically said, *Girl, you are crazy.*

Jon and I had dinner at a new TriBeCa restaurant and spent an exorbitant amount of money on wine. No Barolo this time, but Burgundy and Vouvray. We talked about wine and our jobs, exams and competitions. I asked him about his family and waited patiently for him to ask me about mine. When he didn't, I knew nothing had changed. But I tried to convince myself I could be wrong.

We walked out of the restaurant, two bottles of wine under our belts. Jon grabbed my waist from behind and kissed the back of my neck. Fire shot up and down my body; I felt an electricity return that only Jon could provide. We walked hand in hand to a nearby spirits bar and finished the night with some Calvados.

Jon and I had shared Calvados a few times before, like it was a secret between us. The spirit, a barrel-aged distillate of apple and/or pear, feels both serious and whimsical. I felt grounded in the moment that night, with just enough of a magical aura to question whether it could last. Calvados seems to do this: create and upset reality at the same time.

We went back to Jon's place, even though mine was much closer. I was afraid of what Megan would do to him if she found him in my bed.

I slept well that night. I went home relatively early the next morning, and when I got out of the shower, I had a text from Jon asking if I'd made it home alright. *Maybe he has changed.* I had never before felt that Jon thought about me when I wasn't in his presence.

I didn't text him for a few days, and he didn't make any attempts to get in touch with me either. I texted him late one night.

"Hey. Just thinking about you."

"Hey. I have to work late tonight and be back early tomorrow."

"I wasn't inviting you over. Just saying hi."

"Oh. Hi."

I was prepared to throw the whole thing in the bag again when I got a text from him on Saturday inviting me to the Big Apple BBQ party at Eleven Madison Park on Sunday night. I was supposed to work, but I switched around my schedule so I could go. I was a wreck all day. Anxious. Excited. Scared.

When it was finally time, I rushed to 24th Street and scanned the grand room for Jon. I found him in the bar, an uncontrollable smile breaking out on my face when I saw him. He returned a forced half-smile, but adrenaline was coursing through me at a rate that didn't allow me to be defeated. I moved to embrace him and he gave me a flaccid hug. He introduced me to his brother Jordan, who had come in from New Jersey for the day. Jordan had obviously never heard my name. Jon told me that he was going to take his brother to the bus terminal, and he would come back to meet me.

I knew plenty of people at the party and did my best to be social. I checked my phone every two minutes, afraid of the text that would come about 45 minutes later.

"Hey, I'm really not feeling well. I think I'm going to go home. Can we meet up tomorrow?"

"No, Jon. We need to talk tonight."

"Can't it wait? Tomorrow should be an early night for me."

"No. It shouldn't take long."

He finally conceded and agreed to meet me at Milk & Honey. While I waited, I sat outside Eleven Madison Park on benches put out for the party. Tears streamed down my face. Someone I knew there asked if I was okay. I felt the eyes of many on me, but I didn't care. I thought I was losing Jon forever that night, and it broke my heart.

I arrived before him at Milk & Honey and ordered a Jack Rose: a cocktail made with Applejack, grenadine, and lime. Applejack is an American spirit made from distilled apples, usually combined with vodka or whiskey. It is the American version of Calvados, but without the seriousness or the whimsy. It is exactly what it is. Straightforward, sturdy, easygoing. I was done with ambiguity.

I texted Jon and asked if I could order him anything. He asked for something light. I ordered him a Negroni, which is not light, but I thought the booze could unlock some honesty in him. I just wanted to understand why.

He arrived and sat next to me in a booth. We sipped our drinks. I laid everything on the table.

"I love you, Jon. I think you know that."

His eyes dropped with guilt and remorse.

"And *this* isn't enough. I can't do it anymore."

"Jane. I don't know why I can't be with you. But I can't." My eyes glassed over with tears. He went on, "I know you're the best thing that will ever happen to me, but…I'm not there."

I cried and he consoled me. Our server bought us a round of drinks. Perhaps she knew we were in the industry. Perhaps she felt bad that I was obviously getting dumped.

VIGNETTE

Jon paid the bill, left an enormous tip, and we walked out into the cold together. He asked if he could put me in a cab. He gave me a hug and apologized. For everything.

"I'm not worth you getting this upset over. I'm not that special, Jane."

"To me you are. I hope that means something."

"More than you know."

I cried a lot the next day. And the days to come. But the confusion and anger had lifted. Just sadness remained. I did what I should have done from the beginning: I avoided all interaction with Jon. No tasting groups, no industry events, no dropping by the bar at EMP. I spent my days and my nights studying. A certain clarity had emerged from the emotional wreckage, and I felt strong again.

159

Apples and Pears

Apples get lead billing
when it comes to
Calvados and cider,
but pears provide
a strong supporting
performance. The
following deconstructed
Venn diagram looks
at the similarities and
differences between
the two fruits and the
beverages they create.

JUST APPLES

Sugars	
Minerals	
Appellations (Distillate and Cider)	Fine Bretagne, Cidre Cotentin, Cornouaille, Sidra de Asturias, Sidra Natural de País Vasco
Varieties	Cider and distillate-making apples are split into four categories: sweet, acidic, bittersweet, and bitter (phenolic). Calvados AOP requires vineyards to be planted to a minimum 70% **phenolic** varieties.
Desired Climate	Apples grow better in the cool, wet, coastal regions of Brittany and Asturias than pears do.
Resting Period	
Fermentation	
Stills	

APPLES AND PEARS	JUST PEARS
Fructose, glucose, and sucrose – in both, fructose is the dominant fruit sugar.	Sorbitol is unique to pears, and very hard to ferment with ambient yeast. It is why pear cider is rarely bone dry.
	Both have a lot of minerals, but pears have a slight edge in fiber and potassium.
Calvados, Calvados Domfrontais (min. 30% pear), Calvados Pays d'Auge (max. 30% pear), Eau-de-Vie du Cidre du Maine	Poiré Domfront
	Pears see less of a classification system than apples do. Poiré Domfront AOP has the most specificity in terms of varieties, namely that the white pear tree is recommended over all others.
Normandy is habitable to both pear and apple trees.	The unique inland soils of Domfrontais are quite dry and particularly suited to the pear tree's complex root system.
Both apples and pears are allowed to rest after harvest before pressing. The beginning of an enizmatic breakdown is necessary to unlock the aromatic precursors in pomaceous fruits.	Pears are often left to sit longer, but have a more narrow margin of error. They are less forgiving, and go from unready to rotten pretty quickly.
Both are grated or pulped before pressing.	Pears are also "brewed" before pressing because of the more fibrous and tannic nature of their skins. The pulp is left to settle for at least a day to let **tannins** collect and precipitate out.
Pot still	Column still (spirits that require a sizable amount of pear in the blend tend to favor a column still because it maintains the primary fruit character of the pear)

Beer

162

The more I entered the world of restaurants and food, the more my stomach revolted.

Sharp pains, burning, and bloating were the more glamorous symptoms. It became a constant worry: how my stomach would feel. I drank amaro and tea throughout meals because my stomach was in so much pain. I cancelled plans on occasion because of its volatility. I dreamed of what it would feel like to never have to worry about my digestion.

I didn't know which foods were causing my discomfort. I wasn't so married to any one food that I wouldn't give it up; I just didn't want to change my lifestyle. If my diet was too restrictive, I wouldn't be able to eat out at restaurants, grab food on the go, or have family meal at work. And there continues to be an unfair stigma around food allergies, especially in the restaurant industry. Restrictions are usually met with scorn rather than compassion.

My doctor in New York told me to try giving up gluten.

The look on my face must have been a combination of an eye roll, an upper-lip sneer, and wide-eyed disdain.

"Seriously," she said. "Try it for a week and see how you feel."

So I did. And within four days, I felt miraculously better. All the tension, pain, and discomfort that had been raging in my belly for a decade suddenly eased. It wasn't perfect, but it was monumentally better. For all the years of hoping for an easy solution to my health problems, finally one was here. The trade-off was a non-issue.

VIGNETTE

The hard part was telling people. Restaurants I had frequented had to hear the news that I was now gluten-free. The restaurant where I worked had to make a gluten-free family meal for me. There was some resistance. Some light-hearted ridicule. And I felt incredibly guilty. But, ultimately, life went on. Everyone got over it. And I felt better.

—

I had always wanted to be a beer drinker. I tried. To me, drinking beer was the definition of effortless cool. *Oh, sure, I'll have one more beer. Want to grab a beer after work? We had a couple beers and called it a night.* It's the sommelier's antidote to pretension.

But beer had never really sat well with me. Maybe it was the gluten all along; beer is, at its core, fermented barley and wheat. Or maybe it was that night in the Dominican Republic. Whatever it was, I could never stomach much of it. And now I had an excuse not to drink it.

So for me, beer remains an unexplored expanse. I taste beer for work. I study beer for exams. But it lacks the emotional connection that wine, spirits, and cocktails have provided me. I find emotion in beer's absence: in the gratitude that – sometimes, rarely, and in part – there can be an easy solution.

Classifying Beer

Styles of wine and spirits enjoy a fairly easy classification system: *This is Barolo because it is Nebbiolo made in the Barolo region of Italy. This is bourbon because it's a corn-based distillate aged in charred, new oak.*

Beer is much more complicated. The different styles are defined by a complex matrix of heritage, ingredients, process, and flavor profile. The following chart presents an entry point to looking at beer classification by showing the relative weight of the preceding factors that define each style.

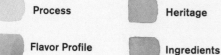

Process

Heritage

Flavor Profile

Ingredients

165

Gose

Bock/Doppelbock

Dortmunder

Old Ale

Brown Ale

Mild Ale

Abbey Dubbel/Trippel/Quad

Bitter

Irish Red Ale

German Pilsner

Scottish Ale

Euro Adjunct Lager

Czech Pilsner

American Wheat Ale

Dunkelweizen

Amber Ale

Pale Ale

Hefeweizen

Witbier

American Adjunct Lager

Scotch Ale

DOMAINE ROSTAING *LA LANDONNE* Côte Rôtie, Rhône Valley, France
HORSEPOWER *SUR ECHALAS VINEYARD SYRAH* Walla Walla, Washington, USA
SADIE FAMILY WINES *COLUMELLA* Swartland, Western Cape, South Africa

Guilty–Displeasure Wine

or SYRAH

In early September 2014, I got
an email that changed everything.

It was from Dustin Wilson, wine director at Eleven
Madison Park.

I had been eyeing the job posting for a new sommelier at
EMP all week. When I thought about where I wanted to work
in New York, the only place that really excited me was Eleven
Madison Park. My celebratory meal there, almost exactly a year
earlier, had been extraordinary. Not just incredible food and wine,
but a riveting experience. It is not as formal as most of the other
three-Michelin-starred restaurants in New York, which appealed to
me. I loved the design, the feel, and the spirit. There was a problem,
of course. Jon.

I pulled up that job posting time and time again, and thought maybe I'd just email Dustin and tell him about my interest, as well as my concerns.

Then it popped up in my inbox. *Looking for a Sommelier.* Dustin was careful in his choice of wording; he is considerate and professional. The last thing he would do is poach someone from another restaurant in the city. But given I had already expressed my interest in moving on, both to Dustin and to my employer, Dustin knew how I would interpret his words. "Think about it," was all he wrote.

Dustin and I had coffee and he laid out the job for me. It didn't sound easy, but it sounded incredibly rewarding. I told him briefly about my health problems, and my concerns there. He didn't pry, but reiterated his support if I wanted to do this.

Dustin told me he knew about me and Jon, and that Jon was excited to potentially work with me. He was confident the two of us could work together professionally. My boss at the time (and good friend) Richard thought Dustin was crazy. "Knowing your and Jon's backstory, I wouldn't hire you if you pissed La Tâche." I didn't urinate $5000 bottles of Burgundy, and I wasn't confident Jon and I could work together professionally. But I agreed to a trial shift at EMP.

I emailed Jon before I began. *I think we should sit down and discuss potentially working together.* He suggested a cocktail. I suggested coffee.

It was the first time Jon and I had talked in almost four months. We had coffee at a small café outside my building, the day before I would trial. The conversation was somewhat illuminating of Jon's psyche. We didn't discuss our past relationship at all. His concerns were purely professional. He was worried I would come in and steal some of his thunder. He didn't know how he'd handle it if I passed the Masters and he didn't.

I was flattered, surprised, and quite amused by the tenor of the conversation. He concluded that, ultimately, he would learn something from me, and that I would be good for the restaurant and the wine program. He told me he was the one who had suggested that Dustin email me. Maybe, I thought, this is what Jon and I were always meant to be: colleagues, friends, maybe even competitors. Maybe this is how we could remain part of each other's lives.

Within a week, I had a job offer. I accepted on the spot.

Dustin invited me to a wine-team dinner the week before I started. It was on a Sunday, and I joined a little late after I got off work. We drank some of the greatest wines in the world: Bartolo Mascarello Barolo, Rene Rostaing Côte Rôtie, Vincent Dauvissat Chablis, all with several decades of age.

The Rostaing was the wine of the night. I have never been a big Syrah drinker (much to the shock and outrage of fellow sommeliers). The aromatics are compelling, but the palate always disappoints me. I find Syrah's structure to be tight, condensed, astringent, and just not that pleasant. But the Rostaing was different. It was open-knit and welcoming, allowing me to parse out each flavor and sensation, drawing out every layer of smoked meat, dried bramble, and crushed black pepper.

The group lingered around the table until only tiny pools remained in each glass. There was talk of going out further. We migrated outside. It was just starting to get cold, October in New York, a gentle nudge for us to decide where to go next. Those who had been eager to join when we were inside slowly peeled off, one by one. The chill in the air, a text from a loved one, an early morning. Whatever it was, Jon and I ended up the only two remaining. We made our way to a SoHo bar called Mother's Ruin, not far from my apartment.

We sat at the bar, my legs squeezed together inside his, and had more drinks than we needed. He told me that ours was "The Great Sommelier Love Story," which I thought was cute, but cheesy and misguided. Jon seemed to have a selective memory for the good times while forgetting all the pain he caused me. Something about him allowed me to participate in this selectivity as well. I remembered the good times. This was one of them. We went home together.

Something felt different this time. Not in him, not in the relationship, but in me. I knew what was important. The job was important to me. My career was important to me. And I'd figure out the rest. I still didn't fully understand my feelings for Jon, but I had spent the last few months proving to myself that I could do without his drug. To have it back was not as sharp in its high or its low.

168

The Flavor Wheels of Syrah

Syrah has a distinctive flavor profile: a combination of tart, jammy, and dried black and blue fruit, black pepper, smoked meat, purple flowers, dried herbs, dark black earth, sometimes an oak presence, and sometimes a presence of **carbonic maceration**. These factors wax and wane in the classic Syrah regions of the world, creating an expression unique to each place. It is hard to create a flavor wheel for Syrah as a grape since these characteristics vary so much in their proportion from region to region. It is useful, however, to see the many wheels of Syrah.

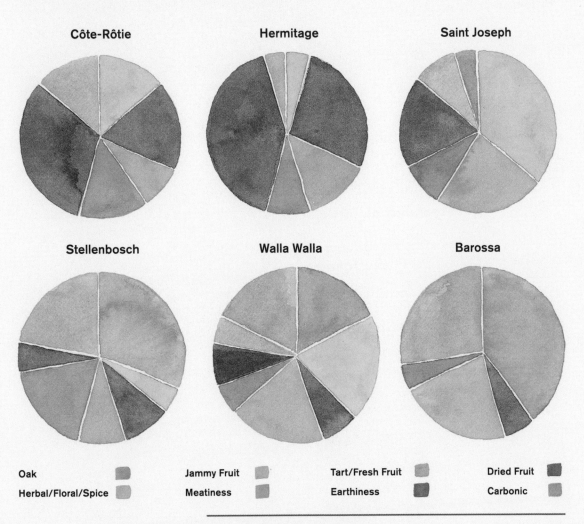

Côte-Rôtie	Hermitage	Saint Joseph
Stellenbosch	Walla Walla	Barossa

Oak

Herbal/Floral/Spice

Jammy Fruit

Meatiness

Tart/Fresh Fruit

Earthiness

Dried Fruit

Carbonic

GUILTY-DISPLEASURE WINE

Somm Survey: What do you generally dislike that the sommelier community embraces?

"I never understood the obsession with the Negroni. It's a bitter, medicinal, gross cocktail."
Pete Bothwell

"Chartreuse. I'm ashamed to admit it, but I have always had a strong aversion to any anise/fennel/wormwood flavor."
Hannah Williams

"Burgundy."
Christopher Bates

"Fernet…seriously, this stuff is disgusting."
Vincent Morrow

"I don't enjoy Nebbiolo as much as the next somm. I have a low bitterness threshold, so the tannin really gets to me."
Chris Ramelb

"I don't freak out about Sake like everyone else does."
Eric S. Crane

"Most low-dosage grower Champagne. Take your $55, two-year-old battery acid walking, son."
Jim Bube

"I'm not a huge fan of cider. The levels of Brettanomyces can sometimes get out of hand, and I don't really enjoy drinking brett-infused bubbly apple (or pear) juice."
Justin Timsit

"I strongly dislike sour beers."
Mia Van de Water

"I cannot stand Fernet-Branca. It tastes like I'm brushing my teeth with tree bark."
Cara Higgins

"Everyone is so hot on the most expensive Burgundies, even when they're not ready to drink. I like wines that are ready to drink. I'm not into killing babies."
Jackson Rohrbaugh

"Champagne. I'm so sorry to say it. Everyone balks, and I lose a fat chunk of street cred when I admit this. But I find the bubbles terribly distracting, not to mention they give me indigestion! However, I LOVE old, flat, room-temperature Champagne, when I think you can actually pick out much more nuance… and the bubbles are usually still present enough to provide a bit of fanciful texture."
Stevie Stacionis

HERMANN J. WIEMER *MAGDALENA VINEYARD RIESLING* Seneca Lake, New York, USA
SHINN ESTATE VINEYARDS *CABERNET FRANC* North Fork of Long Island, New York, USA

New York Wine

The next few weeks revolved around a sole focus: succeeding at Eleven Madison Park.

I hungrily studied the menu, the wine list, and the service manual. I learned the history of the restaurant and the building. I learned every ingredient in every dish. I knew who designed the chairs in our private dining room, who made our trays, and who provided our flower arrangements.

Eleven Madison Park is a New York restaurant. Obviously, it is a restaurant in New York but, more than this, it is a restaurant that attempts to negotiate, translate, and champion the produce and culinary traditions that make the state and city so special. Therefore, the wine of New York plays heavily on the list.

Long Island and the Finger Lakes are the main wine regions of New York, and unlike some other states in the US, they do not have a very illustrious history. Fruit wine, Concord grapes, and Manischewitz were the early stalwarts, with a strong foothold to this day. But a number of determined producers in each region emerged from the rubble to pioneer fine wine. I remember being so inspired by this. To fight for excellence from a place of dearth. *If they can do it*, I thought, *surely I can do this.*

The training to become a sommelier at EMP was intense. Because the knowledge and skills required to be a sommelier are very specific, the restaurant hires for this position, and this position only, from the outside. If you want to be a captain at EMP, you start as a kitchen server and work your way up. It takes most people at least two years. As a sommelier you are fast-tracked, but you are still expected to understand and be able to perform the functions of all the other positions in the restaurant. I took on the weight of this responsibility fully.

The first few weeks were filled with self-doubt. My history was eclectic – wine store manager, craft cocktail bartender, beverage director – but it did not include fine dining experience. I had never worked in a restaurant that had one Michelin star, let alone three. I tried to take everything in, observe people's movements, and imitate them. The first time I was given a plate of food to take to a table, I thought I was going to pass out. I would shake when saucing dishes and tremble while holding trays.

I was a wreck, but a wreck with a goal.

I took Klonopin to get me through this time. I would wake up early in the morning, after going to bed just hours before, with ringing anxiety and despair. I would take Klonopin and ride out these panic attacks, sobbing uncontrollably into my pillow. Eventually, overcome by exhaustion and the pills, I would pass out again. I took this secret with me every day to work. Each night, when I successfully completed a shift, I looked back in disbelief at where I had started the day. The daily evolution was just as monumental as the nightly unraveling that brought me back to tears.

VIGNETTE

Sommelier Feud: New York Wine Edition

We surveyed 100 New York sommeliers, and the top answers are on the board…

In a word, what is the future of New York wine?

Diversity	26	Fun	10
Promising	20	Bright	9
Exciting	17	Strong	4
Amazing	13	Improving	1

Who produces the best New York wine you've ever had?

Hermann J. Wiemer	29	Ravines	6
Red Tail Ridge	19	Bellwether	5
Element Winery	16	Channing Daughters	3
Terrassen	10	Bloomer Creek	3
Dr. Frank	7	Empire Estate	2

What's the best grape being made in the Finger Lakes?

Riesling	86	Cabernet Franc	4
Chardonnay	7	Pinot Noir	3

What's the best grape being made on Long Island?

Cabernet Franc 54
Merlot 21
Sauvignon Blanc 14

Lagrein 6
Blaufränkisch 4
Dornfelder 1

What is the grape you'd like to see winemakers in the Finger Lakes work with/work with more?

Grüner Veltliner 38
Syrah 29
Chenin Blanc 13
Gamay 9
Nebbiolo 5

Sylvaner 3
Petit Rouge 1
Grignet 1
Altesse 1

What is the grape you'd like to see winemakers on Long Island work with/work with more?

Albariño 31
Sémillon 24
Melon de Bourgogne 16
Friulano 12
Fiano 9

Trebbiano 4
Freisa 1
Furmint 1
Pineau d'Aunis 1
Garganega 1

What's the best non-wine product coming out of New York?

Cider 36
Beer 33
Spirits 18
Apples 5
Vermouth 2

Corn 2
Foie Gras 1
Hospitality 1
Cuisine/Chefs 1
Ambition 1

Bordeaux

My 29th birthday was 12 days after I started working at Eleven Madison Park.

At this point, I had not yet begun training as a sommelier at EMP; I was still working my way through the other positions in the restaurant. Jon texted me after I left work and asked what I was doing for my birthday. I told him I was meeting up with a friend, and he invited me out for a drink after. We sat among my new colleagues from EMP and did shots in honor of my birthday. Jon went outside to smoke; I joined him, and we kissed under a sliver of the moon.

When Jon and I had first started dating, when I was lamenting his inability to commit, Jon had told me he couldn't stop thinking about three things: making Eleven Madison Park the best restaurant in the world, becoming a Master Sommelier, and me. I believed him, and I could see this struggle alive and well in him when we started working together. Each of these imperatives was a full-time job, and he couldn't find a way to commit effectively to all three. I was the easiest to drop or, more accurately, not take on.

The commitment I witnessed firsthand from Jon was astounding, even if it wasn't a commitment to me. Jon cared about every detail. On top of the 60 (plus, plus) hours a week he was working, Jon would show up early to study and blind-taste. He always made time to help others out, whether it was a private tasting with a server who was taking their Certified exam or a staff-wide class on pre-Jesus wine (yes, he did that). If you made a mistake, Jon would tell you about it, but it only came from how much he cared.

When I started training as a sommelier, Jon acted professionally and took an active interest in my development. The first day I trained with him was incredibly difficult. He harassed me about every little thing I did, and at the end of the night, we sat down in the cellar and he pulled out a list of the things I had done wrong. The day had been trying on my nerves, both professionally and personally. As I sat in that cellar, listening to him coolly listing all the areas that I needed to improve on, I started crying.

"Should I continue?" he asked.

"I still have feelings for you, Jon. I know you still have feelings for me, too."

I didn't know this. I didn't believe it at all actually, but I wanted him to respond one way or the other. He didn't say a word and left the cellar. He returned a couple minutes later with the bottle I had **Port-tonged** that evening, the first time I had ever done it in service. It was 1988 Château Ducru-Beaucaillou, a second growth from Saint-Julien in Bordeaux.

"I thought you might want to keep this," he said, handing me the bottle.

"Thanks," I said. "You can continue."

Wine–Libs

Fill in the requested words, then transfer them to the text on
the next page to get some "tips" on how to sell, decant, and serve
a bottle of Bordeaux.

1. Direction _____

2. Exclamation _____

    Noun _____

3. Verb _____

    Color _____

    Different color _____

    Adverb _____

4. Verb _____

5. Number _____

    Abstract noun _____

6. Celebrity _____

    First name _____

    Number with four digits _____

    Transitive verb _____

7. Adjective _____

8. Abstract noun _____

    Adverb _____

    Noun _____

9. Noun _____

10. Adjective _____

    Noun _____

11. Body part _____

    Different body part _____

    Piece of clothing _____

    Adverb _____

    Insect _____

    Adverb _____

    Verb _____

12. Transitive verb _____

13. Verb _____

    Body part _____

14. Noun _____

    Verb _____

    Noun _____

15. Sense _____

    Noun _____

    Abstract noun _____

    Census category, plural _____

    Animal, plural _____

16. Piece of furniture _____

    Emotion _____

    Adjective _____

BORDEAUX

Wine–Libs

1. When you approach a table in your restaurant, it is important to always stand to the _____ [direction] of the host.

2. You can start the conversation by saying " _____ [exclamation], can I be of any assistance with the wine _____ [noun]?"

3. Try to ascertain what the guest wants, with questions like: "What sort of wine do you normally _____ [verb]? Do you feel like _____ [color] wine or _____ [different color] wine tonight? Are there any grapes that you _____ [adverb] dislike?"

4. Make sure you know what the guest will be eating, in case you're asked what would _____ [verb] well with their meal.

5. Try to figure out what price range to make recommendations in. Depending on the guests, you can simply just state a number (i.e. "Is something around $ _____ [number] what you're looking for?") or point to a few different price points to gauge their _____ [abstract noun].

6. Take the order, repeating back to the host the name of the producer, bottling (if applicable), and vintage, i.e. "Excellent, I'll go get for you the Château _____ [celebrity] Saint- _____ [first name] from the _____ [number with four digits] vintage." Offer to _____ [transitive verb] the wine list away.

7. Set _____ [adjective] glasses for each of the guests.

8. Make sure you handle the bottle with the utmost _____ [abstract noun]. If it needs to be decanted, place it _____ [adverb] in a cradle as to not disturb the _____ [noun].

9. When presenting the bottle, repeat back the producer, bottling, and vintage. Wait for a _____ [noun] from the host that it is the bottle they ordered.

10. If decanting tableside, set up a **gueridon** with the _____ [adjective] tools you need. This includes several coasters, torchons, a wine key, a decanter, a candle, and don't forget the most important tool of all, a _____ [noun].

11. Wheel the gueridon over, with the bottle still in its cradle. Cut the foil at the second _____ [body part]. You will have to flip your _____ [different body part] to cut all the way around without moving the bottle. Remove the foil and put it in your _____ [piece of clothing]. Wipe down the cork with a torchon. _____ [adverb] drill the _____ [insect] of your corkscrew into the cork. Just as _____ [adverb], pull the cork out. Use your torchon to _____ [verb] down the bottle again.

12. _____ [transitive verb] the cork to the host.

13. Light the candle. Being careful not to _____ [verb] the sediment, pick up the bottle and begin to pour the wine from the bottle into the decanter. The candle should be right below the _____ [body part] of the bottle, allowing you to see any sediment that might be approaching the top.

14. Pour as much wine as you can, while not allowing the _____ [noun] to flow into the decanter. Now you can _____ [verb] down the bottle, which may have accumulated _____ [noun] on its surface while aging in the cellar.

15. Pour a _____ [sense] of the wine for the host. When they approve, move _____ [noun]-wise around the table, pouring the guest of _____ [abstract noun] first, then the _____ [census category, plural], then the _____ [animal, plural].

16. Leave the bottle and the decanter on coasters at the _____ [piece of furniture]. Watch in _____ [emotion] as the _____ [adjective] guests enjoy the spoils of your wine service!

Port

There is a purpose to the Port-tong, beyond it being a good show.

Vintage Port is made to age in bottle for upwards of 30 years. Port ages under cork, which maintains its shape and integrity when held within the confines of the glass bottleneck. But when you try to disrupt the cork's slumber, it often revolts, crumbling and breaking off with every torque of the wine key.

Port tongs were invented in Portugal's Douro Valley in the 18th century. The premise is to open the bottle without disturbing the cork. The tongs are two c-shaped pokers that form a circle when pressed together. They are heated over an open flame and then applied to the neck of the bottle, just below where the cork ends. (Don't do this at home. If you do, make sure you remove any paper adhesives or metal capsules before applying the tongs.)

The tongs are held for about ten seconds, then shifted, so the heat is applied evenly. (The c-shapes don't quite meet each other, and there is always a small gap that goes untouched on the first application.) Another five seconds does the trick. The next step – the change in temperature – is what cracks the glass. At Eleven Madison Park, we used a barber's brush dipped in an ice-bath. Outside the dining room, we'd also been known to use water guns, liquid nitrogen, and just throwing a glass of ice water at the bottle (the latter is especially effective when Port-tonging Champagne).

The cracking noise is subtle, so when the cork comes off easily, still within its glass casing, the audience is always surprised and delighted. At EMP, it was rare for a table to order a full bottle of Port, so we'd use this trick to open any older bottles of wine (or whenever we thought the guest could use a good show). Microscopic glass shards can break off, so we'd always decant the wine through a mesh sieve. This has the dual benefit of separating the wine from its solids.

Port-tonging is a true meeting of function and form. It serves a purpose while offering a beautiful and engaging presentation. It also got Port, once viewed as a stodgy drink of the British elite, back on the lips (literally and figuratively) of New York City sommeliers.

180

Quantifying a Vineyard Site

Many regions seek to establish a hierarchy for their vineyard sites. Most of the time, these hierarchies are based on a wisdom-of-the-elders sort of pedagogy. *We don't know exactly who made these decisions or how they came to them, but these are the best sites.* Empirical knowledge over the years can question these models – you'll hear things like, "This **Premier Cru** is practically **Grand Cru** quality," and "This **Grosse Lage** site underperforms" – but it doesn't tend to change classifications much.

Enter the *beneficio* of Porto. Each vineyard is given points based on 12 different factors, which ultimately determine how much wine is authorized to be fortified from that vineyard. (*Beneficio* both describes the classification system and the act of **fortification** in Porto.) When the system was conceived in 1947, it was acclaimed for its ability to prevent the market from being saturated with Port as well as to keep the quality of Port quite high. Today, the IVDP (Instituto dos Vinhos do Douro e Porto) scores and classifies the vineyards of the Douro every year, controlling the production volume as well as the price a grower can receive for their grapes for Port production.

Today, many Port producers are critical of the system, claiming that it is used to subsidize non-fortified production in the Douro, that it favors volume over quality, and that it has failed to stabilize the market. Some recommend a reform; others, a complete abolishment. Whatever its current shortcomings, the scale (detailed on the following pages) provides a unique and quantitative insight into the inner workings of a vineyard classification system. The elders have spoken.

Quantifying a Vineyard Site

Location

The Douro is split into five sections, with a number of sub-sections within each, and specific points allotted to each sub-section.
Minimum score -50   **Maximum score** 600

Altitude

Lower altitudes are prized because it has historically been too cold at higher ones. Altitudes over 650 meters receive -900 points!
Minimum score -900   **Maximum score** 240

Slope

The higher the slope, the better, as it increases sun exposure and reflection off rivers.
Minimum score 1   **Maximum score** 101

Bedrock

Schist gets all of the positive points; anything else is in the negative, including -250 for granite and -400 for alluvium.
Minimum score -400   **Maximum score** 100

Rough Matter

This is a measure of the texture of the soils. Stonier or rougher soils allow rainwater to penetrate, reflect sunlight, and act as a heat reserve.
Minimum score 0   **Maximum score** 80

Exposure

In the cooler western areas, southerly exposure is preferred. In the warmer regions, a northern exposure is preferred.
Minimum score -30   **Maximum score** 100

Shelter

The more sheltered sites, on the tributaries of the Douro River, are more prized.
Minimum score 0   **Maximum score** 60

Yield

This is based on the idea that the more productive a vineyard is, the poorer the wines. 120 points is granted to all vineyards under 55 hectoliters per hectare.
Minimum score 0   **Maximum score** 120

Vinestock

Positive scores are given for the highest amount of recommended grapes (29 recommended out of over 100 authorized), as well as a diverse planting of different grapes as an insurance policy against disease and inclement weather.
Minimum score -150   **Maximum score** 150

Planting Density

Higher density reduces the vigor of each vine, producing higher quality grapes at lower yields. Every vineyard planted above 4000 vines per hectare gets the full 50 points (there are exceptions for vineyards planted before 1998 or for some specific lower density vine training systems).
Minimum score 0   **Maximum score** 50

Training System

Vines that are trained closer to the ground yield riper fruit and earn more points. Pergolas are not allowed in Port production.
Minimum score 0   **Maximum score** 100

Age

Vines less than five years old are excluded from Port production. After that, the scale of points is based on vine age and in general rewards older vines.
Minimum score 0   **Maximum score** 60

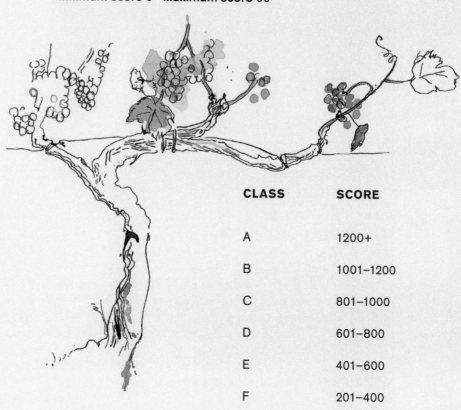

CLASS	SCORE
A	1200+
B	1001–1200
C	801–1000
D	601–800
E	401–600
F	201–400

PORT

More Barolo

I settled into my role at Eleven
Madison Park.

I loved the job, I did well at it, and I seemed to be accepted by
the team. Jon and I continued to see each other. No real dates, no
discussion of emotions, just late nights and early mornings. Again.
Jon seemed to have deep affection for me in some moments and
couldn't look me in the eye in others. I didn't understand why there
was a switch, or how it was flipped.

During this period, I started dating someone else. Perhaps the
first real candidate for something serious since I met Jon. He worked
for a hedge fund in SoHo, and we met at a friend's birthday party.
He had a well-kept one-bedroom apartment down the street from my
own, appointed with beautiful mid-century furniture, a record player,
and a well-stocked bar. He thought it was cool that I used to bartend
at The Violet Hour and that I worked at Eleven Madison Park.

I basked in his adoration. He took me out on a first date that
didn't end with us naked. It ended with us listening to Radiohead
on vinyl and making out on his living-room rug. We discussed *House
of Leaves* and *Invisible Cities* into the early hours of the morning.
He'd had problems with anxiety and could empathize with what
I was going through. On paper, he was perfect. And we got along
great. He didn't make my body hum the way Jon did, though, and
I wondered if this was a good thing or a bad thing.

At the beginning of December 2014, I got a tattoo. I was off that day and took a different train home from running errands. I walked past an art gallery and peered inside. I saw a tattoo parlor in the back. I walked in. I was greeted by a man with, not surprisingly, a lot of tattoos. I told him that I was interested in getting one.

"When?"

"Any availability tonight?"

He laughed. "Sure, come back in an hour."

I did. They gave me a shot of bourbon and took to my ribs. It hurt, but I have a strong tolerance for pain, both in the emotional and the physical realm. The tattoo is a zipper across my ribcage that slowly fades out as the teeth extend toward my back. Something about being closer to what was going on inside my body.

Hedge Fund (Megan's not-so-inventive name for the new guy) came over later that night. He found it thrilling that I got a tattoo on a whim.

I told Jon the next day that I'd gotten a tattoo. "I know, I saw," he said, referencing my Instagram post. "Had you been thinking about that for a while?"

"No, it was sort of a whim."

"Yeah, seems like it," he said, without any tone of playful jest, and walked away.

He spent the rest of service avoiding me. When we did cellar work at the end of the night, he barely grunted in my direction. ENOUGH, I thought.

"You don't have to treat me like this," I confronted him. "We can just be colleagues."

He stood there, blank. I made myself clear: "It's over."

Jon seemed shocked and hurt; I'd never seen him like this. He staggered away, muttering, "Yeah, sure, whatever."

When I got home that night I needed one more ounce of closure. I drafted a text to him that looked like it was meant for someone else. Totally normal.

"I told Jon it's over. I still love him. But I can't do this anymore. Things are going well with Hedge Fund guy. I'd rather put my energy in there, and not be hurt by Jon anymore."

I kept the text on my screen for a moment, debating whether to send or not. I wanted him to know that I still loved him. I wanted him to understand that this wasn't me not caring, it was me caring too much. I hated myself a little bit in this moment, but I sent it anyway. I followed it up immediately with a pretty convincing (at least in my mind), "Well...fuck me. Obviously that wasn't meant for you."

I waited to see his response. I'd never been happier to see the three pulsing dots on my screen. They seemed to dance forever before his message finally popped up.

"Jane, we should talk. I want to say that my feelings for you have grown now that we see each other so much more often. But I feel like that's just screwing with your head. I was thinking while I was away that it would be a good idea to not see each other so we don't get ourselves into trouble again. But then when we tasted on Monday, I realized that I don't want that."

—

Jon and I both worked the next day. He was gentler and more attentive, though we didn't talk about the night before. I left work and got a text from him once I was in a cab, asking if I wanted to meet up for a drink. We went back to Mother's Ruin and sat side by side in a booth in the back.

We made small talk about work before he unloaded everything he had to say to me. And it was everything I'd ever dreamed Jon would say to me.

"I think about you all the time. I think about us ending up together, and it makes me happy. I think about how I will introduce you to my parents. I want us to be together."

I reminded him of our past. And I told him that I didn't trust him.

His head hung down and he grabbed my hand. "I know I have to earn your trust, and I know I've done a lot of things to abuse that trust in the past. I'm ready to fight for you."

I stifled tears at these last words. Partly tears of happiness, hearing what I'd always wanted to hear from Jon. Partly tears of sorrow, doubting that he would actually follow through.

I told him I was seeing someone else. And that I would continue to see him until I was given a good reason not to. Jon walked me home. I didn't invite him up, and he didn't ask. He gave me a kiss on the cheek and wished me sweet dreams.

The next morning Jon and I attended a blind-tasting group. I called California Viognier as Vouvray and he called Sutter Home Moscato as German Riesling. A pathetic showing, for sure. As we sat next to each other, I nervously wrote a note on a piece of paper and slid it to him: "My afternoon appointment has been cancelled, would you like to have lunch?"

We walked through TriBeCa and ended up at a sun-drenched 187 West Village restaurant. He ordered a half bottle of Alain Graillot Crozes-Hermitage and we ate lunch. He didn't have to work that night and drank most of the bottle; he was already tipsy, smiley, cute. He walked me home. I asked him something about his niece, and he asked me about mine.

"I don't have a niece, Jon."

"Oh, I thought you did. Your sister..."

"Is married. No kids."

It was in this moment that Jon realized he knew nothing about me. That he'd never asked. And from then on, he started asking.

—

That Saturday, I had a date with Hedge Fund. Jon knew. He'd asked me what I was doing on my night off and I crinkled my face, as if to say, *Do you really want to know?* I got a text from Jon in the middle of the evening. "Can I heckle you while you're on a date?" I smiled.

Hedge Fund and I had a nice night. Dinner, drinks, then back to his place. He asked me to spend the night. I told him I'd rather sleep in my bed. It was 2am. I walked home and checked my phone. I had several texts from Jon.

How was your night?

What are you doing?

Do I want to know...?

Come meet up with me. I've thought about you all day.

We met at Churchill's, our after-work neighborhood bar. I settled in, said hi to some work friends, and got a drink. He stood by my side.

"You're beautiful," he said, a few Guinnesses and shots of Jameson in.

"That's the first time you've ever told me that."

"I know."

188

—

We were both scheduled off the following Thursday, perhaps a plot by Dustin, who had joked about Jon and me getting married since my first day at Eleven Madison Park. Jon asked if he could take me on a date. "Our first real date," he said.

We went to a two-Michelin-starred restaurant with an open kitchen and acted like teenagers; we couldn't keep our hands off each other. I had to work lunch the next day so I was slower with the alcohol. Jon is a lightweight and was drunk by our third course. By the time we got back to my place, I was practically pouring him into bed. He got very serious, in the hilarious way that only drunk people can.

"JANE. I do love you, I do, I do. I love you…"

His words trailed off into a drunken snore. I sat beside his passed-out body and cried. Sure, he was drunk. But I knew he meant it.

—

Jon soberly told me he loved me within the week. And within a few weeks after that, we said it to each other every day. I broke things off with Hedge Fund.

Jon had plans to go home to New Jersey for Christmas. We celebrated together on the night of the 23rd, before he departed. I was off from work and cooked him dinner at his place. We drank a white wine from Franz Hirtzberger, a rare blend of Pinot Blanc and Pinot Gris from 1994. For red wine, we drank a bottle of our beloved Cordero di Montezemolo. 1971 vintage.

I thought about the first time Jon and I had Barolo together. And the first time he served it to me at Eleven Madison Park. This time, I had a deeper understanding of the wine. My love for it didn't come from a place of naïve exuberance, but from knowledge and respect. I now understood the soil this wine was grown in, the large oak botte that raised it, and how remarkable it was that over 40 years after these grapes were grown, it tasted the way it did. My emergent emotional intelligence also allowed the wine to affect me differently. It's not that everything is a miracle when drinking Nebbiolo. It's that everything seems like one. And that is enough.

VIGNETTE

Somm Survey: What wine has been part of a love story in your life?

"Salon Champagne. It is the wine I bribed my wife with to agree to marry me!"
Desmond Echavarrie

"Immich-Batterieberg. My wife and I met making wine there."
Christopher Bates

"Sherry for sure. My other half, Carla, wooed me with it before hitting me over the head with a bottle in Prospect Park."
Richard Rza Betts

"Sherry!"
Carla Rza Betts

"Krug Grande Cuvée. The first time my husband ever surprised me with wine, this was it."
Christy Fuhrman

"My fiancé brought a bottle of 2001 Olga Raffault Chinon to our first date. This random tech guy from a dating app was suddenly a lot more interesting, and we drank down the bottle while closing the restaurant talking."
Claire Hill

"Lustau Fino Sherry was the wine Adrienne and I were drinking when we realized our relationship was going to be more than platonic."
Eric Hastings

"Cordero di Montezemolo."
Jonathan Ross

"A long time ago, I was on a date with a young man and I let him pick the wine. He impressed me when he selected a Mâcon Viré-Clessé – it was perfect with the meal, and not too expensive. But then the next date he ordered something terrible. I was confused so I asked him where he learned about wine. He said, 'Oh, I don't actually know anything about wine, I just always order the one with the most accent marks.'"
Kelli White

"My husband and I got engaged in Clos Saint Jacques in Gevrey-Chambertin. It gave us a good excuse to collect a bunch of it while we were there."
Mia Van de Water

"Jean Foillard Morgon Côte du Py. I first had it on a date with Ryan. It was early in our relationship, we were celebrating my birthday, and it was the first time I ever drank a bottle of Cru Beaujolais. I loved every sip."
Rebecca Flynn

"La Tour Vieille Collioure Rouge. My husband and I had it on our first date. It is a deceptively simple wine, a country red. When he pulled this out for dinner instead of a big-name brand to try to impress me, I knew he was the one."
Victoria James

"My true love is simply working the floor with my corkscrew and opening bottles. I love it so much."
Yannick Benjamin

C.V.N.E. *IMPERIAL GRAN RESERVA* Rioja, Spain
RIDGE *EAST BENCH VINEYARD ZINFANDEL*
Dry Creek Valley, Sonoma County, California, USA
PENFOLDS *GRANGE* South Australia, Australia

Guilty–Pleasure Wine

or AMERICAN-OAKED WINE

During this time, things started to get easier with my health.

I could go for drinks with my new colleagues after work and sleep through the night after. I stopped getting panic attacks, and the fear of my body gradually lessened. The gauze was still there, but it was soft and thin, just reminding me ever so gently that something wasn't right. The punishing and painful character that had been so stubborn during my withdrawal gradually faded away.

I will remember these as some of the best months of my life; I was so happy and grateful to be alive and (somewhat) well. I was excelling at my new job, opening some of the greatest wine in the world alongside some of the greatest food. I adored my co-workers and had already built a strong community inside the restaurant. My love for Jon was finally reciprocated, and we settled into a happy and loving relationship.

We had many conversations about what had transpired in our less-than-illustrious past. He was overly apologetic and never gave me any reason to doubt his current sincerity and commitment.

But I couldn't get past the need to understand why. I craved understanding, not atonement.

And, as with most things, it was not a simple or straightforward explanation, but a complex constellation of factors. He had recently gotten out of a long and difficult relationship when we met. He wanted to be single. He wanted to focus on his career. I had come on strong. I was intense. He was afraid of what falling for me meant.

I had to accept that it was not one thing that had created the initial conditions. It was the turn of a slope, the gradient in soil, hail at flowering, a heat wave in mid-August, a windy September, and finally, rain at harvest. But, despite early estimations, everything had settled well and was looking harmonious. Sometimes a rocky vintage makes the best wines.

—

One morning in mid-January, I awoke to a venom I had not experienced in months. I had thought it was gone. In those first few moments of wakefulness, my awareness unfolded in layers. I realized I was awake, and then I realized I was in the midst of a panic attack, and then I realized what this meant.

I was devastated. I sobbed on the pillow until I woke Jon up. He held me, convulsing and trembling, until my eyes had no more tears and my body lost its will to shake. He had no idea what was going on or what to do about it. This was a side of me he hadn't seen before.

Jon proved his devotion in the months to come as I struggled to readjust to being ill. I knew it was a horrible position to put him in: watching the person you love being held hostage to an invisible disease and unable to do anything about it. He didn't always say or do the right thing. He sometimes left my side too soon or stayed too long. His affirmations of *you can do anything* did not do justice to what I was enduring. I projected onto him my impatience with my ailments, and figured he was sick of hearing about it and dealing with me. I did not make it easy for him, but he stuck with it. Stuck with me. He never made me feel like a burden.

I was also incredibly, inescapably jealous of him. What a gift – to go through life and not fear one's body. To go to bed not fearing how you would feel when you woke up. To make plans for the future and be genuinely excited about them, not afraid of how you will feel when they arrive. When I couldn't sleep, I would resent Jon's restful body lying next to me.

GUILTY-PLEASURE WINE

Despite my resentment of his wellness, and despite my tumultuous and difficult temperament, Jon didn't flinch. He held me tight through my fits of panic. He worked shifts for me at EMP when I was feeling too poorly. He even let me take the wheel in deciding what we would drink – no small concession in a sommelier relationship.

I didn't feel up to drinking much in this time, but when I did, I developed a strange craving for wines that were drenched in American oak (which is not that cool for a sommelier). Rioja. Aussie Shiraz. Some US Zinfandel. These wines were rife with notes of coconut, dill, and vanilla. They reminded me of bourbon (which I certainly didn't feel well enough to drink), and I felt that if ever there was a time to indulge in a guilty pleasure, it was now.

—

While my illness brought me and Jon closer together, it drew me away from my ambition. The theory portion of the Master Sommelier exam was approaching, and I was a mess. I never woke up except mid-panic attack, and the adrenaline coursed through my body through the rest of the day. I couldn't focus well enough to study for any significant amount of time, and I was taking Klonopin again to find relief.

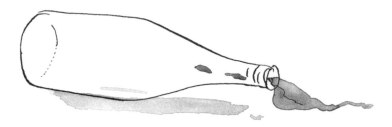

Somm Survey:
Is there an alcoholic beverage you consider a guilty pleasure?

"Tequila. I drink it in victory and in defeat. I drink it in happiness and in sorrow. I drink it when I try to be poetic, like right now."
Justin Timsit

"I fully embrace white wine spritzers."
Rebecca Flynn

"Lindemans Framboise."
Christopher Bates

"Gin and tonic, Spanish-style, served in a Burgundy glass with enough gin to put you in a very good mood after the first one."
Benjamin Hasko

"I love a good Miami Vice."
Cedric Nicaise

"Moscato d'Asti."
Caleb Ganzer

"I sometimes love big full-bodied Bordeaux blends from New World regions."
Jeffrey Porter

"Sauvignon Blanc in all forms (preference to barrel fermented/aged), Highlife and Jäger."
Theo Lieberman

"Oaky California Chardonnay."
Christy Fuhrman

"Aperol Spritz – it especially becomes a very big problem during the summer. The bottles of Champagne in my house are replaced by Prosecco during this time of the year."
Paula de Pano

"To quote a friend, 'I never feel guilty about drinking.'"
Eric S. Crane

"Budweiser."
Eric Hastings

"A big Shiraz with greasy burgers and fries. Something about the alcohol level mixed with the ketchup, fat, and salt makes it such a good pairing."
Jackson Rohrbaugh

"A really well-made, tart Cosmopolitan. Ideally shared with my mom."
Kelli White

The Oak Equation

Why oak? Oak, compared to other types of wood and other aging vessels, has the benefits of being tight-grained (no leaking!), having some porosity (wine is exposed to some oxygen and experiences development), as well as (when new) augmenting color, increasing **tannin**, and contributing (often desired) flavors of vanilla, baking spice, and cedar.

1. Amount of New Oak

Oxygen Ingress
without Flavor
Influence

Oak-Derived
Flavors Galore

0%   100%

2. Size of Barrel

In general, the smaller the oak barrel, the more the oak influence (with less wine in the barrel, there is a greater surface-area-to-wine ratio).

Feuillette
132 liters

Barrique
225 liters

Hogshead
300 liters

Puncheon
400–500
liters

Demi-Muid
600 liters

Stück
1200 liters

**Foudre/
Botte**
2000–
10,000
liters

3. Type of Oak

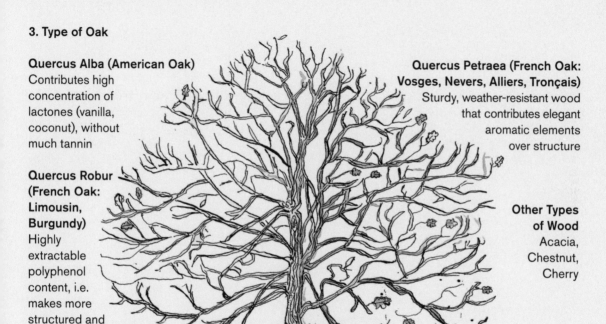

Quercus Alba (American Oak)
Contributes high concentration of lactones (vanilla, coconut), without much tannin

Quercus Robur (French Oak: Limousin, Burgundy)
Highly extractable polyphenol content, i.e. makes more structured and less aromatic wines

Quercus Petraea (French Oak: Vosges, Nevers, Alliers, Tronçais)
Sturdy, weather-resistant wood that contributes elegant aromatic elements over structure

Other Types of Wood
Acacia, Chestnut, Cherry

4. Alternate Aging Vessels

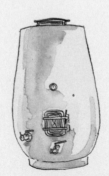

Ceramic Egg   **Concrete Vats**   **Clay Amphora**   **Stainless Steel**   **Glass Demi-Johns**

Amarone

During this time I had a very specific dream:

Mona is not a city anyone would willingly choose to live in. Yet once there – either by duress or happenstance – it is hard to bring oneself to leave. There is a strange comfort to it.

Mona is a city of sadness. The stone streets are polished with the tears of its inhabitants. The men on the trains tell stories of burned-down houses, sick babies, and cold nights. Some feel the pain and the sadness more acutely than others, but every inhabitant is righteous in the individuality and severity of their pain, believing the person next to them could never understand it. More pain is continually caused by this bombastic self-involvement, and the inhabitants drift further and further apart in their concentric circles.

I am one who comes and goes from Mona more freely than most. I am not a foreigner when I am there; I come often enough and stay long enough that I am recognized as an inhabitant, and thus suitably disdained. But unlike many, I do not take comfort in our mutual affliction. I look people in the eye. I pick up a dropped hat. I try to smile through the tears. I try to laugh at the absurdity of my sadness. I look for beauty in our communal anguish. I find bits, pieces, threads of joy.

I am, however, righteously convinced that the suffering endured by those inside Mona is far greater than what anyone on the outside could experience. No acute or situational pain can compare with our chronic misery. There is a hierarchy of pain.

VIGNETTE

Each night I am in Mona, I throw a dinner party and invite all its inhabitants. I serve radicchio and horse meat and open bottles of "the great bitter one." I always think, *maybe this time someone else will come,* but they never do. I drink Amarone alone and tuck the leftovers away.

AMARONE

A Raisin's Reason

Amarone della Valpolicella is a **DOCG** in the Veneto region of Italy. Its name means "the great bitter one." Its production methods make it a singular style of wine: quite high in alcohol, flush with dried-fruit flavors, packing a perceptible amount of **residual sugar**, and mountainous in structure (both in terms of acidity and **tannin**). It is a behemoth of a wine but manages to find balance for its fruit and alcohol through sweetness, structure, and – of course – bitterness.

The grape aging process, known as *appassimento*, is what makes Amarone so unique. After grapes are harvested, they are laid out to dehydrate, either on straw mats or in small crates. This process lasts on average between three and four months, and in this time the grapes change considerably. These changes are what give Amarone its specific and unique characteristics.

 A grape shrinks to half its weight due to water evaporation.

 The sugar content rises to 25–30% as the grape dehydrates.

 Botrytis, or the noble rot, can sometimes develop. Botrytis can enhance glycerin in the finished wine, but is often avoided for fear of degrading the skins and introducing unwanted flavors.

 Polyphenols are concentrated.

 Fructose content rises almost four times more than glucose content; fructose is a sweeter-tasting sugar than glucose.

 Acidity decreases: tartaric acid by about one-third and malic acid by over two-thirds.

 Reservatrol is produced in significant quantities, a compound that has been cited as one of the healthiest in wine production.

VIGNETTE

Irish Whiskey

Competition season started
up again in early 2015.

I didn't shake (much) in everyday service at the restaurant, but
in particularly nerve-wracking situations, the tremors returned.
I developed some tricks for shaking less. Choking up on a bottle
or decanter, rather than holding it from the base, alleviated some
of the shaking. Since the Copenhagen competition, I would say
in my head, *let them see you shake*, but was actually thinking, *god,
no, please don't let me shake*. So I clenched tight, tried to steady
myself, and just prayed they wouldn't ask me to double decant.

Double decanting is the practice of decanting a bottle of
wine into a decanter, and then pouring the decanted wine back
into the bottle. The reasons can be many: you can double decant
for sediment, in which case you need to rinse out the bottle before
pouring the wine back in. And you can "splash" double decant,
in which aeration is the primary concern, and there is no sediment.
The latter is a tidy task for a competition as it requires no extra
supplies to dump or rinse and can easily be done table-side. Many
restaurants keep a funnel on hand to assist in double decanting,
but the more virtuosic practice – and certainly the one required
at competitions and exams – is to free pour the wine from the
decanter back into the bottle. This requires a steady hand and
supreme focus.

Top New Somm was right before the Masters exam in 2015, and it was being filmed for the TV show. My theory, I hoped, would be strong, as it was just two months before the Masters. I had not been tasting much, so I didn't have high expectations for my performance on the tasting portion. I was most nervous about service, though. My service skills on the floor of a restaurant had always been strong, and my service score on the Advanced exam was excellent. But since the Klonopin withdrawal, my body was unable to handle the stress. I was worried about shaking. The cycle of anxiety perpetuated itself, and the more I became nervous about anxiety, the more anxious I was.

The day of the competition, no surprise, I was a wreck. Adrenaline coursed through my body in waves, in what felt like one long, drawn-out panic attack.

I had been doing more research on the science of Klonopin withdrawal. Klonopin binds to something called a GABA receptor – a receptor in the brain that is responsible for us feeling calm. When the body is accustomed to a benzodiazepine, it can no longer bind to those receptors without it. When the daily presence of the drug is removed, the receptors go through a long and erratic healing process. And as they heal, they essentially offer no protection against the ravages of normal and daily stress (not to mention the added stressors I was subjecting myself to). I lived in a constant state of adrenaline high, my body unable to feel peace or contentment. When I experienced elevated levels of stress, I would get dizzy, start sweating, feel heart palpitations, and shake. I felt like I was swimming in a thick pool of air, unable to breathe properly or see straight.

My body was also irrationally sensitive to all stimuli. A piece of silverware being dropped in the dining room would cause a blast in my body, shaking me from the core out with a thousand pinpricks. My alarm going off in the morning would jolt through my nervous system, causing my heart to pulse incessantly in my limbs. I was even more sensitive to touch and found myself recoiling from even the gentlest caress as though it were a hot poker.

All these sensations were elevated when I felt stress, but I fought as hard as I could to suppress them that morning in February. Twelve extremely talented East Coast-based sommeliers, many of them my friends, convened at Corkbuzz on 13th Street at 8am. Most of us had worked late the night before, but everyone was happy to be there and happy to see one another. The community that was built through these competitions was strong. There were very few contentious attitudes, only support and camaraderie.

I sat there wishing I could enjoy it more than I did, but the tide of panic wouldn't let me feel any sense of ease.

Before I went into service, I sat down for an OTC (off-the-cuff in television lingo) interview with the producers. They asked if there was any topic I hoped I wouldn't encounter in the service portion.

"Sake? Beer? Whiskey?" they asked.

I shook my head. "No, I'm ready for anything."

"Anything you hope you don't have to do, service-wise?"

I paused.

"Double decanting," I said. "I've been shaky lately, and I worry that I would not have a steady enough hand to do it well."

—

I navigated the beginning of the service competition with dread. I answered questions I knew and crafted pairing menus that should have been fun. But I couldn't enjoy it. I was too afraid of what would come next. I managed to affect a natural, breezy demeanor even though I was screaming inside.

"Great, we'll take the Quintarelli," one judge said. "Can you go ahead and double decant that at the table?"

Instant panic. The producers of the TV show were watching the live reel in the next room, no doubt elated by this happy coincidence: what great television.

I placed down a wine glass for each guest, as well as three coasters, and rested the decanter on one of them. I brought over the bottle and presented it to the host. I opened the bottle and only started to shake slightly as I poured the wine from the bottle into the decanter. This, of course, was the easy part. As I switched hands and began pouring the decanted wine back into the bottle, I started to shake violently. My hand ratcheted up and down as I attempted to steady the stream of wine back into the bottle. I knew one small slip-up would mean streams of wine down the side of the bottle and a pool on the floor.

Usually in exam or competition decanting, the judges will continue to engage the contestant during decanting. *What grapes are in this wine? What was this vintage like? Who else makes wine like this?* The judges were too stunned – or maybe they didn't want to make it any more difficult for me – and sat there in silence for

IRISH WHISKEY

what felt like the longest 30 seconds of my life. I was so mentally and physically exhausted that I didn't think I would finish decanting. I thought one of my wrists would give out, or my legs would collapse under me, or maybe I would pass out altogether. But I kept pushing through, only because I didn't know what else to do.

Miraculously, I finished the decant. And even more miraculously, I didn't spill a single drop. This is a rare feat for any double decant, let alone one that involved a steadiness more typical of an epileptic than a sommelier.

I wondered how many points I lost at the table. Technically, I didn't do anything wrong. But if style points exist, I should have had a negative number. To me, the job of the sommelier is to make the guests feel comfortable about wine service. *How comfortable would people feel to have a visibly nervous sommelier?*

I moved on to the next service table, relieved to no longer be staring those two men in the face and hopeful that this table *had to be better.* And it was. The judges asked about a sparkling Tasmanian wine I was familiar with. I served it, my body limp in relief for an easier task than double decanting. They asked me questions about whisk(e)y – American, Irish, and Scotch. This was and is my wheelhouse, and my energy and passion shone through. I offered more information than they asked for, but in a digestible, service-oriented manner. When I left the table to prepare the whiskey service on the side station, I heard one judge say to another, "Who is this girl?"

After it was over, we all went to a bar next door to wait for results. Jon and I had to go to work later that day, so I wasn't going to order anything – but then I saw the Redbreast on the back bar. Something about Irish whiskey seemed suddenly perfect. I was grateful for the moment of ease and purpose it had granted me in the competition; moreover, the flavor profile was one of utmost comfort. With less peat, less malted barley, and greater distillations than Scotch, Irish pot-stilled whiskey is honeyed and plump. The sharp edges that make Scotch interesting are replaced by smooth lines that make Irish whiskey comforting. I got a few odd looks; my neat shot of Redbreast was quite the contrast to everyone else's pink and orange-strewn Negronis.

In announcing the results, the judges said the winner of Top New Somm came down to two candidates: me and Morgan Harris, another New York City sommelier who was featured on the television show.

"The winner this year gave an exceptional tasting, which really pushed them over the edge."

Definitely not me.

"Morgan Harris!"

I clapped genuinely and openly. I was relieved it was over, and happy I did as well as I did. I certainly didn't deserve to win given how I had tasted. Furthermore, I was happy for Morgan. I was pretty sure I would get the wild card spot for Top New Somm and would be competing in the nationals. But before I could worry about nationals, I had to focus on my main concern: theory for the Masters was approaching in less than two months.

Wines Defined by Economic Necessity

We'd like to think that all great wines (and spirits) are products of artistry. That there is a creative genius somewhere who is devoted to constructing the purest and most terroir-driven style, with little consideration of other factors. But vignerons do not act in a vacuum, and many iconic styles of wine (and spirits) have been created out of economic necessity rather than artistic license.

Single Pot Still Irish Whiskey

Whiskey has been made in Ireland for centuries. The country's archetypal style – single pot still (formerly called pure pot still) – came about in the late 18th century in response to a tax levied on the use of malted barley. Producers turned to unmalted barley instead. Though the tax was later repealed, the style stuck. Its current legal definition requires the use of a minimum 30% each of unmalted and malted barley, as well as distillation in a pot still.

Rioja

The Tempranillo-based wines of Rioja, Spain, enjoy a long and rustic history. It wasn't until the mid-19th century, due to influence from the Bordelais, that the wines of Rioja began to be aged in cask. It is unclear why American oak was chosen instead of French. The best theories are that Spain already had strong trade routes established with the US, that French oak was less plentiful, and that American oak was cheaper. No one posits that this was a creative choice. It was a choice that stemmed from necessity and ease, and one that has defined the wines of Rioja to this day.

Tokaji

Several legends of economic necessity surround the creation of **Tokaji**, the great sweet wine of northeastern Hungary. The style was "discovered" in the late 16th or early 17th century. Supposedly, farmers were called off to battle the Ottomans right before harvest began. When they returned to their vineyards, the grapes had shriveled on the vine. Unable to afford losing the harvest, the farmers picked them anyway and made the world's first **botrytis**-influenced wine. When these wines took years to ferment, only yielded an alcohol of 3%, and required intense vineyard labor, economic necessity stepped in again, and the fermenting must of non-botrytized grapes was added. This style – of combining heavily botrytized and non-botrytized grapes – is what defines the iconic Tokaji Aszú today.

Non-Vintage Champagne

Traditionally, Champagne was always a blend of multiple vintages. It wasn't that Champagne houses believed blending to be the pinnacle of expression: it was an economic survival technique. The climate of Champagne is on the cusp of being too cold for grape growing. Certain vintages, historically even more so than now, were disastrous and yielded little good fruit. To stave off the economic impact of these hard years, the government required houses to keep a reserve of wines from previous years. To this day, the government controls how much of these reserve wines is released. Multiple vintages are blended together to create a house style that is the same from year to year, a consistency that proves to be another economic boon for the region.

Grappa

Italians have a history of being plucky, resourceful, and poor. In the Testaccio neighborhood of Rome, they invented nose-to-tail dining, called *quinto quarto*, where you can sample lungs, nerves, intestines, and other delicacies. Grappa is the quinto quarto of the wine world. A product that would otherwise go to waste – the leftover skins (called pomace) of a wine fermentation – is instead watered down and distilled. Grappa historically allowed poor farmers to make an extra buck on their harvest, as well as providing a tranquilizing warmth to subsistence living.

IRISH WHISKEY

Eau de Vie

The next few months drew out in a crescendo of gnawing anxiety.

I was trying to study as much as I could, but also trying to get rest and take care of myself. I couldn't decide what I wanted more – to feel good or to pass the Masters – and I couldn't figure out how to achieve either. But as far as my co-workers at Eleven Madison Park and the TV cameras were concerned, everything was fine.

The day of the Masters theory exam, I woke up at 5am with a surge of adrenaline through my body. With the frequency of these wake-up calls, you'd think I would be used to them, but every time it happened was uncomfortable and upsetting. I took a Klonopin to get back to sleep. I tossed and turned for another 45 minutes, plagued by irrational thoughts of blanking on the easiest of questions, MS proctors laughing at me, and the repercussions of failing. I took another Klonopin, silent tears streaming down my face. I was mad at myself for taking the drugs again, and mad that I had ever taken them to begin with. I lay with these self-recriminations until, finally, the heavy sheath of drugs knocked me out.

I woke up at 10am. The panic attack had abated, slaughtered by the strength of the drugs. It was replaced by a thick haze that made everything heavy, distant, and fuzzy. My body hummed from my core. This hum seemed to be a product of my anxiety combined with drug-related withdrawal/relapse symptoms.

VIGNETTE

I held everything in as tight as I could to prevent these feelings from surfacing.

That morning, before the exam, the TV crew came to my room. They filmed me getting ready, pinning my Advanced pin to my jacket, and talking about my final thoughts and anxieties. They were nice people just doing their job, but I felt like telling them all to *fuck off and leave me alone.*

When I got in the room for the theory exam, the shaking was so severe that I couldn't let my arms hang loosely by my sides. I clenched throughout the exam and felt very little relief when I left the room. I knew I had not passed. There were questions I missed because the drum of anxiety had clouded my thinking. There were questions I missed because the haze of medication had clouded my judgement. But, when it came down to it, my knowledge was not where it needed to be to pass. I was not a Master.

Jon and I went out that night. He had also not passed theory that day. We had reservations at a fancy restaurant in Atlanta. *To celebrate,* we had thought. I resurrected a tradition that I had learned with my parents when we were in Prague in 2011: a shot of pálinka before dinner. In Budapest, the locals use this to whet the appetite and celebrate the breaking of bread. We bastardized it, literally drowning our sorrows with some pure, hard alcohol in our empty stomachs.

I could have celebrated this as a moment in the journey – if only I felt better. That both my health and my MS progress were in such disrepair was hard to manage. I wondered whether these goals were incompatible, and whether I would ultimately have to choose.

Eau de Vie Traditions

Kirschwasser
Germanic Countries

Kirchwasser is unaged cherry brandy. It is often served chilled as an aperitif with a side of water, though some higher-quality examples are served at room temperature. Kirsch is also quite common in cooking and is used in the traditional German Black Forest cake.

Slivovitz
Eastern Europe

Slivovitz is the name for plum eau de vies in Eastern Europe. Its name is said to have come directly from the Polish word for Passover, Święto Paschy. Slivovitz gained the seal of Kosher for Passover and is thus regulated to only contain sugar, water, yeast, and plums. It is drunk both before and after dinner, as well as at all special occasions and rites of passage.

Pálinka
Hungary

While the urbanites in Budapest drink pálinka before dinner, the Hungarian Romanis have the strangest traditions with the spirit: only drinking it upon waking, at a funeral, or as preparation for a garbage-scavenging trip.

Rakija
Serbia

Rakija is the umbrella category for fruit-based brandies in Serbia (Šljivovica is the variety made with plums). Rakija is commonly homemade in Serbia. Everyone either makes some or has a friend or cousin who does. It is rude to refuse an offering of Rakija. It is served in a shot glass but meant to be sipped.

CLOS ROUGEARD *LES POYEUX* Saumur-Champigny, Loire Valley, France
DOMAINE DE BELLIVIÈRE *ROUGE-GORGE* Coteaux du Loir, Loire Valley, France
PUZELAT-BONHOMME *KO IN CÔT WE TRUST* Touraine, Loire Valley, France

Loire Valley Reds

The weeks that followed the exam, I expected some level of relief.

At least it is over. However, my body didn't register this fact and I continued to suffer. If I had passed theory, at least it would have all been for something, I felt. As weeks and months passed and I didn't return to studying or tasting, I knew I would not sit for my Masters in 2016. Emotionally, physically, and mentally, I was wholly unprepared to embark on the process again.

—

By the time of the nationals for Top New Somm, I had no desire to compete. All I wanted was to not think about exams and competitions and my ability (or lack thereof) to succeed in them. This attitude seemed to help me in some respects: I slept better without the help of Klonopin, I didn't feel as nervous, and I didn't shake much during service (this last fact was also helped by the wondrous discovery of beta blockers). But, with the nerves, I also lost some of my edge.

LOIRE VALLEY REDS

My excitability – the passion and purpose I'd shown at both the EMP competition and TopSomm regionals – was gone.

The competition went by with little event. I imagine my service was quite forgettable. My tasting was weak. And as evidenced by the theory portion, all the knowledge I had crammed so tightly into my head two months before had slowly begun seeping out. I gave a lackluster performance, and – not surprisingly – did not win.

2015 wasn't my year, but it was Jon's. He competed in TopSomm (the over-30 age bracket). He felt good about his tasting and theory, but thought maybe he had flubbed a portion of service. Jon was a wizard when it came to service. If he had forgotten a step or missed a beat, I was confident the judges wouldn't have noticed (or cared). It was about how Jon made people feel when he served them. Jon's guests always felt they were in the company of greatness, and that this greatness existed to serve them. I had witnessed it as a guest and as a colleague. It never failed to stop me in my tracks.

The awards ceremony was held at a local restaurant. In announcing TopSomm, the judges named the first and second runners-up – neither of whom were Jon. Then Geoff Kruth started describing the winner.

"The winner this year is great at what sommeliers need to be great at: service."

At that point, I knew Jon had won, and I was so proud of him.

We spent the rest of the evening celebrating with our peers and the Master Sommeliers who had judged the competition. The MS who had proctored Jon's service portion told him he had never given anyone a perfect score on service – until today. That same MS also told me that I was a "pussy" for considering not sitting the exam next year.

I left that night with a series of overwhelming emotions: joy for Jon's success, frustration at my own stagnation, and confusion about what to do next. Jon and I stayed in Napa the next few days and found ways to continue celebrating (and to blow his prize money). We drank old Dunn Howell Mountain Cabernet, old Stony Hill Gewürztraminer, and young Domaine de l'Arlot Romanée-Saint-Vivant. We did a vertical tasting of older Vérité. We drank Friulian orange wine with new friends.

But the wines that had the greatest impression on me that week were wines I didn't even drink. In the service portion of the competition, we had been asked to name several (I think it was five) red wines from five different grapes grown in the Loire Valley. Cabernet Franc is the most famous red grape of the Loire, but Pinot Noir, Gamay, Côt (aka Malbec), Pineau d'Aunis, Cabernet

Sauvignon, Pinot Meunier, and Grolleau also come into play. In this (relatively) small stretch of about 1000 kilometers, these diverse grapes present a study on many of the flavors and processes of the wine world: **pyrazines, reduction,** oak, **carbonic maceration, rotundone**. Though this was a service question, it was a lesson I needed to learn in theory and in tasting. I didn't *really* understand these wines – nor did I have the attention to pay them just then. I knew the key was there, though. That when I was ready to examine and understand why and how those wines tasted the way they did, I would be back on track.

211

Flavors of Loire Valley Reds

	TECHNICAL TERM	RED GRAPES OF THE LOIRE FOUND IN
White and Black Pepper, Smoke	Rotundone	Pineau d'Aunis
Green Bell Peppers, Green Bean	Pyrazines	Cabernet Franc and Cabernet Sauvignon
Candied and Pickled Red Fruit	Carbonic maceration	Gamay, Côt, and Cabernet Franc
Smoke, Gunflint, Burnt Rubber	Reduction	Cabernet Franc
Vanilla, Toast, Cedar	Oak influence	Cabernet Franc, Cabernet Sauvignon, Pinot Noir

WHY	WHERE	PRODUCER TO TRY
Aromatic compound forms during fermentation because of precursor amino acids present in grape juice/skins.	Coteaux de la Loir	Domaine de Bellivière
Aromatic compound forms during fermentation because of precursor amino acids present in grape juice/skins.	Chinon, Bourgueil, Saumur-Champigny (most notably)	Olga Raffault
Without being crushed, whole berries begin fermentation inside their skins. The grapes are in a sealed container, and the CO_2 created (and sometimes added) keeps the vessel oxygen-free.	A stylistic choice seen throughout the Loire	Puzelat-Bonhomme
A complex concept, **reduction** refers to the result of reductive winemaking techniques that produce a reduced wine (confused, anyone?). Basically, through limited exposure to oxygen and maximum exposure to solids (and other factors, including fermentation temperature, soil composition, and grape variety), winemakers can encourage the formation of sulfides.	Chinon (most notably)	Bernard Baudry
While the tradition of maturing wine in large, old casks that impart less flavor is still alive and well in the Loire, some of the region's top producers are pushing boundaries by maturing their top wines in smaller oak barrels, some of which are new and impart more flavor.	Saumur-Champigny, Chinon, Sancerre	Clos Rougeard

LOIRE VALLEY REDS

DOMAINE ECONOMOU *OIKONOMOY LIATIKO* Crete, Greece
HATZIDAKIS WINERY *ASSYRTIKO DE MYLOS* Santorini, Greece

Greek Wine

Back in New York, after TopSomm,
I tried to be a normal twenty-something
and enjoy my life.

I didn't go to tasting groups or study. I focused on work at Eleven
Madison Park and tried not to think about my body. I slept in late
and caught up with my friends. I went out after work and drank
whiskey and ate cheesy fries. I didn't go to doctors' offices or chase
promised cures.

In a sense, this was freeing. So much mental energy had gone
into figuring out what was wrong with me and how to make it better.
Freeing up this mental space was liberating, but the bad days
haunted me. *Could I just accept that this is how I would feel, for
the rest of my life?*

I couldn't, not yet. Late summer my symptoms started to
intensify. I had more bad days where my body rang so loud and
so insistently that I couldn't think about anything else.

I took some solace in work on these days, knowing that even
if I could not make myself happy, I could make other people happy.
Eleven Madison Park allowed people to be outside of themselves
for an evening. Many times, this was celebrating a birthday or
anniversary with loved ones. Sometimes it was a business meeting
or recognition of a career achievement. Other times it was a cancer
patient's bucket-list meal. Or a woman's first meal out after caring

for her sick parents. We allowed people to punctuate their lives on a daily basis, and I found great comfort in being a part of this.

—

After several months of struggling, and many teary phone calls, my mom insisted I call the Mayo Clinic. This medical fortress in Rochester, Minnesota, had always been on my parents' radar as a place I should go if I couldn't find answers anywhere else. I had resisted going, for several reasons. I had seen dozens of doctors over the years, in some of the best hospitals in New York and Chicago. I couldn't believe there was something all these doctors had missed or overlooked. I thought maybe the solution lay in an approach that wasn't strictly Western, and the Mayo Clinic's scope did not include alternative medicine. Going to Mayo also felt, in some ways, like the end of the road. If they couldn't help me, could anyone?

For these reasons, it was an emotional decision to go to Mayo. But at my parents' insistence, I agreed. I took off a week in October. I told an abbreviated version of the truth to my boss Cedric, the wine director of Eleven Madison Park after Dustin. I told a couple of my best friends at work but kept it from most people. To minimize the lying, I told people I was visiting family in Minnesota. This was technically true: my mom would be at the clinic with me.

She picked me up from the airport Monday morning in Minneapolis, and we drove about 90 minutes to Rochester. It was early October and already cold and windy in Minnesota. We caught up about the family and our lives, but there was a somber undertone to the drive.

She had booked us suites down the hall from one another in a modest hotel in Rochester. It turned out to be connected to the Mayo Clinic by a labyrinth of underground passageways. We could have easily spent the week there and never ventured outside. Malls, restaurants, and all the medical facilities were connected.

As I walked into the first appointment that afternoon, I was already emotional. The personal and medical hardships of the past years bubbled below the surface, needing only light provocation to erupt. The intake appointment was performed by a professional and kind young doctor. I found a parallel between what she and I did daily: I am expected to give each and every guest a once-in-a-lifetime experience. The expectations are fiercely high. These doctors see patients every day who have suffered for years and have come to them for answers; expectations are similarly towering.

As I described what I had gone through and my symptoms, I started to tear up. The doctor handed me a tissue and we continued. At the end of the intake appointment, she offered a hypothesis.

"We have many appointments scheduled for you this week. My opinion is that we will rule out any definitive medical cause of your symptoms. Your anxiety, your sensitivity to stimulus, your chronic fatigue and pain – all point me to believe that the cause of this is a sort of fibromyalgia called Central Sensitization Syndrome."

Central Sensitization Syndrome. It sounded very official. I lit up at the idea that there could be a diagnosis at the end of all this. That there could be a name for what I was going through. *If there's a name for it, there has to be a treatment, right?*

Unlike a normal vacation full of fun activities that I would want to feel good for, I didn't put any pressure on myself to feel good this week. It was my week to be sick and to be treated like I was sick. While the rest of my life depended on me ignoring this identity, this week I could be honest about what I was going through. And, perhaps as a result, I felt better. I woke up clear and calm. My mom and I watched movies and ordered room service. We found the best cocktail bar Rochester had to offer, and we frequented a Greek restaurant down the block from our hotel.

The restaurant had great gluten-free bread and an impressive selection of humble Greek wines. We sampled a number throughout the week: Assyrtiko, Xinomavro, Liatiko, Moschofilero, Savatiano. None was more than $12 for a glass, yet all were clean and seemed to be carefully made. It occurred to me that the Greeks were responsible for developing wine traditions all over Europe, yet – due to occupation, war, and disease – had fallen behind in fine-wine production. Like ungrateful children all over the continent, winegrowers in Burgundy, Montalcino, and Tokaj forgot about this important piece of their history.

I remember scrolling through Instagram that week and discovering that a sommelier friend was on vacation in Burgundy. Pictures of DRC, Coche-Dury, and Roulot popped up, all from excellent vineyards and vintages. (An expensive trip, for sure.) I didn't think my Greek wines would have quite the same social-media cachet, but I wouldn't have traded the peace in Minnesota for all the Périgord truffles in the world.

I thanked my mom for the week. I had been an ungrateful child at points in my life.

—

The diagnosis came as predicted: Central Sensitization Disorder.

"Well, how do we treat that?" I asked eagerly. I figured there was a specific pill for this particular brand of fibromyalgia. *Bring it on!*

"A combination of therapy and psychiatric medicine." My heart sunk. "We can't tell you which will work for you, but you should find a good psychiatrist and therapist in New York, and they will be able to help you."

At that point, I had tried almost every psychiatric drug on the market as well as a few different types of therapy. In and out of doctors' offices, on and off medications, was the story of my life for the last ten years. And none of them had been able to help. I was back at the beginning. I realized then what "Central Sensitization Syndrome" was. It was a diagnosis for people who couldn't be diagnosed. It was a gesture toward helping those who couldn't be helped.

Historical Vineyards of Greece

Without the Greeks, wine would not be as we know it. Greece was one of the most influential early wine cultures, shaping the viticultural traditions of most of Europe (and thus, the world). Due to hardships of war, disease, politics, and economy, Greece fell behind other fine-wine regions in modern times. But Greece was making fine wine – and singing epic poems about it – before most other regions had even been discovered. Here are a few of the ancient wines of Greece, and the lore that surround them.

Maronitis Oenos

Maronia, in southern Thrace, is mentioned by Homer as the favorite wine of his hero Ulysses. Ulysses uses Maronitis Oenos to poison (i.e. get drunk) the Cyclops Polyphemus, who couldn't stop drinking the sweet, near-black wine.

Ariousios Oenos

Ariousios Oenos, on the island of Chios, was prized by the connoisseurs of antiquity for its fine dry wines. In 1822, the island was burned from end to end by the Ottomans, and the viticulture there has yet to recover.

Siatista

The vineyards of Siatista, in western Macedonia, were widely esteemed during Ottoman rule. **Phylloxera** devastated this area in the late 19th century, though there have been recent efforts to revive the vineyards. Siatista was well known for its straw wines – where grapes are dried on straw mats to dehydrate and concentrate the sugars – which were produced from the little-known red grape Moschomavro.

PATRICK BOTTEX *LA CUEILLE* Bugey Cerdon, Savoie, France
CLETO CHIARLI *VIGNETO CIALDINI* Lambrusco Grasparossa di Castelvetro, Emilia-Romagna, Italy
GIACOMO BOLOGNA *BRAIDA* Brachetto d'Acqui, Piedmont, Italy

Breakfast Wine

Much of my life was in line with what
I expected it would be as I turned 30.

I not only had a job I loved, but a successful career. I had incredible
friends, a loving family, and a wonderful boyfriend. But I couldn't
reconcile my 30-year-old self with such crippling health problems.
If someone had told my 22-year-old self that things would be as bad
as ever when I turned 30, would I have had the strength to go on?
All the strange and beautiful things that happened along the way, and
the hope that someday I could find relief, were what kept me going.

—

The months after the Mayo Clinic proceeded with some normalcy.
I was functional and continued to perform well at work. I went to
see new therapists and psychiatrists, trying new medications and
new relaxation techniques. This process, which I had been through
before, was always exhausting. It required a vigilance over my body
that was all-consuming and wholly unpleasant. All I wanted was to
not think about my body, and here I was examining it with an internal
microscope every time I changed medication or felt worse.

*Is this new dosage working? Do I feel worse because of the
medication, or was it something I ate, drank, or did? Is this*

therapist working, or should I try a new technique? Would acupuncture help? How much is reasonable to pay for my treatments? How much can I afford? Am I still experiencing remnants of Klonopin withdrawal, or am I back to the baseline? Would I feel better if I didn't drink alcohol at all, or would this just deprive me of one of my great pleasures?

Ad nauseum, on repeat, inside my head. I continued to be consumed by the thought of getting back to the Masters exam. I sat out in 2016, choosing not to take theory. I stood by and watched as Jon and many of my friends passed theory in March. I was so proud of Jon, yet I couldn't help but feel a little sad for myself.

The first thing Jon said to me after he took it was, "You would have passed." I realized he was trying to affirm my knowledge and talents, but it just made me feel foolish for not sitting. In May, when Jon passed his service portion, I felt the pang all the more deeply. This time, though, it was more because I still wasn't in any kind of shape to be taking a service exam. The shakiness, the nerves, the dizziness, the anxiety – none of it was going away. I couldn't imagine taking that exam without passing out, let alone actually passing it.

In moments like these, I thought about what lesson I should be taking to heart. What actions would I be able to look back on and say, *I did the right thing*? Is the lesson that I can do anything I put my mind to? That no matter the obstacle, I can overcome it to achieve my goal? Or is the lesson that I should be good to myself, take care of myself, and respect my limitations?

—

In an effort to get back on the wine horse, I agreed to attend a conference in Texas. It had been over a year – from when I took my Masters theory exam – since I had done anything in the wine world. No tastings, no classes, no competitions, no exams. I had excised myself from the community completely. I missed it, but I was also nervous about re-entering, answering questions about my absence, and questioning my ability to perform at the level I once did.

The conference in Texas was a wine show. I wasn't judging the wines, but had been invited to write reviews of those that received medals. It seemed like a gentle, toe-dipping sort of way to get back in the game.

The first evening was fun. Sommeliers from all over the country poured in. People I'd competed against, took exams beside, and gone on wine trips with. A few too many Negronis were had, but I couldn't bring myself to leave and go to bed early. I was so excited to be back with my friends, back in the community, and perhaps back on track to take my exams.

The next day came early. We met in a large conference room to introduce ourselves and discuss the next few days. There were several dozen Master Sommeliers in the room as well as many other respected authorities on wine from all over the world. They made up the judges panel. The sommelier panel, of which I was a part, included many Advanced and Certified sommeliers.

I entered the room, grabbed some fruit and juice, and sat down at one of the dozen or so large round tables. David Keck, one of the organizers of the event and a friend, came to say good morning.

"Hey, we're going to pour some bubbles for the group in a minute, can I enlist your help?"

I froze. I hadn't brought the beta blockers, the medication that had helped me stop the shaking at TopSomm nationals.

"Yeah, sure, of course."

I started to panic. I sat silently while my hot blood and last night's booze coursed up and down my body. *I'm going to shake in front of all these people.* What could I do? I thought about feigning sickness and heading to the bathroom, but I knew they'd all just think I had partied too hard the previous night. This was not the reputation I wanted to develop. I couldn't think of another excuse, another way out. So I would pour.

It's just Cava. At breakfast. I served thousand-dollar bottles of Champagne to some of the biggest collectors in New York – how could I be nervous about this? I mustered a meager *let them see you shake,* the mantra that had worked in Copenhagen, but I was not very convincing.

I started to pour. I choked up on the bottle to have better control. A few glasses were fine. Then my wrist moved funny or the bubbles rose too fast, and panic shot through my body. I started to shake. I attempted to mask the shaking by moving the bottle, switching the position of my hand, reaching over with my left hand and holding my right elbow while I poured. I saw a few people doing double takes, obviously aware of my shaking or strange posture.

I poured for probably only two minutes, but it felt like the whole morning had been spent in that inferno. By the time we got to the venue for the wine competition, I was crashing from my adrenaline rush. I was exhausted and limp. My body felt like jelly, only sustained and upright through the splinter of pounding anxiety that coursed through me. Tears fought themselves up, a salty pinch that lay just below the surface, reminding me how little control I had.

Just then, I got a phone call from my sister. I excused myself to take it.

Beth had also had a hard weekend. She'd had a miscarriage. She was going back to work that day and didn't know how to behave.

"I don't want to pretend I'm cheery and act like everything is okay. But I don't particularly want to tell them what's going on."

I told her I understood. And that in my experience, there were a few things she could do. I told her that sometimes putting on a happy face, smiling, laughing, and talking to other people can get you out of a funk. It's hard to make yourself do it, but it can work.

"And option B, when you really can't bring yourself to pretend that things are okay, is to allow people to see the pain, but use a more comfortable story about why you are hurting. Tell them you're having a hard time with your latest directing project. Tell them you got some sad news in your family."

We hung up. I looked at the group of sommeliers who were opening and decanting bottles for the day's tastings, demonstrating proper technique for the local volunteers. The thought of going back over there made my stomach turn, and I knew I would shake again if I had to double decant.

I quietly slipped out of the room to the huge, 20-stall restroom. I tucked myself in a back stall and finally let the tears come out. I couldn't stop crying and I couldn't catch my breath. When someone came in the bathroom, I held everything in until they left, then collapsed back into tears.

It wasn't just the shaking at breakfast that I was so desperate over. It was what it meant to me. And my future. *How could I entertain taking the Masters again if I couldn't even pour Cava for a toast?*

I took my own advice and texted David that I'd had some hard news in my family and just needed a minute. It wasn't a complete lie – I had had some hard news in my family – but it wasn't the real reason I couldn't get myself off the floor of the bathroom right then.

At the time I couldn't even comprehend how disrespectful and insensitive this was to Beth's suffering. But back then, I still believed that there was a hierarchy of pain, and that mine was supreme.

I texted Jon, who sent me back a picture of himself and Bo. "It's just breakfast wine," he texted.

222

—

I've never been much of a day drinker.

I feel like there's a countdown on the amount of time I can spend drinking before I fall asleep, or before I do something stupid. In Chicago, I was known for having parties that I would only see the beginning of, silently departing to my room to sleep while guests were still in my home. Or, on the other end of the spectrum, one fateful fourth of July, I ended up drinking all afternoon, pushing past the fatigue, only to end the night by playing beer pong with Fireball and puking all over Jon's parents' house.

But there's something to be said for breakfast wine (or perhaps more appropriately, brunch wine). Brunch wine feels like it shouldn't be that serious. You don't have to swirl brunch wine or talk about its legs. It just needs to be delicious, easy to drink, and have a celebratory feel. It should also be, at least for my purposes, delicate in alcohol content. No one wants to stumble out of brunch. Or fall asleep face down in it. And, most importantly, no one should stress over how it's poured.

Wines for Brunching

The wines profiled here are traditional styles made in Italy and France. They use indigenous grapes, and have a distinct flavor and feel that speak to how and where they are made.

They are also all appropriate for brunch. They are serious styles of wine that feel frivolous, several are lighter in alcohol, and many have some sweetness to them. All are various shades of pink. And all have bubbles.

	BUGEY CERDON
Color	Pink (rosé)
Bubbles	Legal minimum of 3 bars of pressure (considered fully sparkling), but often doesn't reach the fizziness of Champagne (5–6 bars)
Alcohol	Lower than standard table wine, usually 7.5–8.5%
Sugar	Legal minimum of 40 grams per liter of **residual sugar** – some definite sweetness!
Regions Producing	Bugey Cerdon is a legal designation of the AOP Bugey in Savoie, France
Grape(s)	A blend of the red grapes Gamay and Poulsard
Best Producers	Patrick Bottex, Renardat-Fâche
Infamous Producer	None!
Breakfast Pairing	Cheese and strawberry blintzes

LAMBRUSCO	BRACHETTO
Pink or red (rosato o rosso)	Called rosso by Italian law, but actually more pink in color
Can be frizzante (lightly sparkling, 1–2.5 bars of pressure) or spumante (fully sparkling, 3+ bars)	Traditionally under 2 bars of pressure (lightly sparkling), but spumante styles may also be made (as well as the rare still wine)
Typically in the range of standard table wine, 11–12.5%	Lowest of all, 5–6% alcohol on average
Classic rosso spumante is usually dry, but semi-secco and dolce styles may be made	No minimum residual sugar requirement for Brachetto, but in practice it is medium sweet (but still refreshing)
Several DOCs throughout Emilia Romagna as well as one in Lombardy	**DOCG** of Brachetto d'Acqui is in Piedmont, where the only significant plantings of the grape are
Many different varieties of the Lambrusco grape exist: Grasparossa, Sorbara, Salamino, Montovano	Brachetto! (Up to 3% other grapes are allowed in Brachetto d'Acqui)
Cleto Chiarli, Lini 910, Cantina di Sorbara, Vignetto Saetti	Giacomo Bologna, Malvira, Marenco
Riunite	Banfi Rosa Regale
Poached eggs, roasted tomato, polenta, and prosciutto	Nutella and raspberry French toast

Boxed Wine (A Crossword Puzzle)

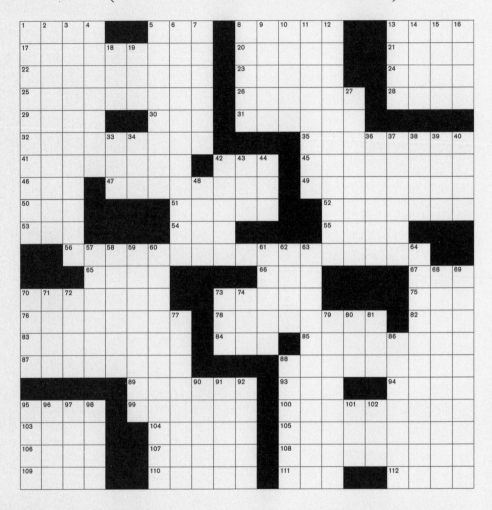

What's the best brunch activity? With a glass of light, sparkling, pink wine in hand, what else but a crossword puzzle!

ACROSS

1. An important village of the Wachau, bckwds.
5. "Hardy har ____ "
8. A Halloween creature, scrambled
13. Part of a wine bottle
17. Wine lover
20. Côte ____ , phonetically
21. "I'm (practically) an open book" (in text message)
22. **Karen MacNeil's seminal text**
23. The sixth page to fill out
24. Napa's Pritchard ____
25. ____ Foothills ____
26. Part of the county that houses Edna Valley AVA
28. An important lake in Canadian viticulture
29. Keystrokes in an online true-or-false test

VIGNETTE

30. To be, in Spanish
31. Favorite card game in the US Midwest
32. Australian Riesling in tinnies
35. Take out on the town
41. Where you'd find the sweetest German styles
42. Best type of wood for aging wine
45. Route between Pays Nantais and Cognac
46. Young boy
47. California producer whose name means "the coasts"
49. What to do with dirty laundry
50. Period in time
51. Epic poem that describes the "wine-dark" sea
52. Rise to Honor, et al.
53. Footwear retailer
54. Adele's "Million Years ____ "
55. Trashy and classy, in short
56. Akin to an old dog learning new tricks
65. The _____ , an Australian passenger train between Adelaide and Darwin
66. Town on the Goulburn Valley Highway
67. A witch, scrambled
70. Pop group Everything but _____
73. She shares a first name with my cat
75. Accusatory texting
76. City with restaurant Elizabeth On 37th
78. "I'm stuck in a _____ . I can't drink anything but Muscadet and Chinon."
82. Early 2000s Irish band, in acronym
83. Many a Pokémon character
84. American business magazine
85. Reports
87. Popular American eating and drinking magazine
88. White
89. Nickname for Paltrow
93. Adelaide _____ , abbr.
94. A US state's best young singers, in acronym
95. The boxed word in this puzzle
99. A bad day

100. "Calm down, crazy"
103. Mount _____ Vineyards
104. Most Master Sommeliers' personalities
105. Online booze finder
106. Vines are not propagated through this
107. God of the winds, responsible for the name of one of the sub-AVAs of Willamette Valley
108. Winter locale
109. Bits
110. What you can do to the day
111. German lady pig
112. Uber _____

DOWN
1. Cloudy
2. Where the world's best cocktail talent is?
3. Sicilian romance novel title, perhaps
4. Taking less force to stay moving
5. Greeting to the soils of Lanzarote
6. What the DOCs Valpolicella, Soave, and Prosecco have in common
7. Puts a part back on Mr. Potato Head
8. Friend _____
9. Nickname for a famous chapel in Vienna
10. Seneca and Cato were this
11. Characterized by piano hits of the 1920s
12. Its synonyms include Tinto Roriz and Ull de Llebre
13. Dönnhoff's anbaugebiete
14. Title used in many Arab countries
15. US state with the most AVAs, colloquially
16. It's good for you
18. Award-winning Midwestern beer, as it is commonly known
19. Ending of island mentioned in 64-down
27. Wine rebellion
33. Nada
34. The first in the franchise was set in Las Vegas
36. Tasti-_____ in formal writing
37. _____ oriented

38. " _____ therefore I am": missive after a particularly drunk night
39. Austrian art museum in New York City
40. Creatures in *Lord of the Rings*
42. Sean Thackery wine
43. The American equivalent of Australia's RAC
44. Child
48. Late winemaker Raffault
57. _____ bacon sandwich
58. An annoying sequel: *A Fish Called* _____
59. Luck into what's rightfully yours
60. Nagambie Lakes, for example
61. Second label of Etude
62. Occasional vineyard pest
63. Famous dessert at Eleven Madison Park
64. Bay of Naples, poetically
68. Exclamation to Chapel Down's region
69. Heavy metal sommeliers?
70. Devil worshiper's week ending remark
71. City in Rioja
72. Rachel Ray staple, in acronym
73. 551
74. A very long time
77. Religious or romantic sentiment
79. One of the owners of Clos St. Jacques
80. Home of the Trojans, in acronym
81. US agency founded in November of 2001
86. 0.404686 hectares _____ (three words, and not totally grammatically correct – sorry!)
88. Taffy ambition
90. Not full, in Catalan
91. Listlessness, scrambled
92. They eat until they die
95. _____ side Road, Russian River
96. Lightbulb
97. Require
98. Bin _____ (last few bottles)
101. Where _____ ?
102. Airport code in Auckland

Mezcal

Jon and I found a week to get away that spring.

We rented an apartment in Tulum, about two hours south of Cancún on the Yucatan Peninsula. Neither of us had spent much time in Mexico, and we were excited about the pristine beaches, cenotes, mezcal, and strong yoga culture (more me for the latter).

Our first day in town, we rode bikes to the beach. It was sweltering, and we were dripping with sweat by the time we got there. We found a humble beachside hotel and café and ordered some mezcal and a few snacks. The mezcal was served in shot glasses, and on the side were wedges of orange and what looked like some sort of spiced salt. It was delicious: the mixture of sweet, sour, spicy, and salty with the smokiness and strength of the mezcal. (I later learned this salt was *sal de gusano*, a mixture of salt, ground chiles, and roasted ground worms.)

Though tequila was still off-limits for me – reminding me of that fateful night in New York where shots flowed freely – mezcal felt like an entirely different animal. I was still hesitant about going too hard with mezcal, but my body didn't feel the visceral suspicion it did with tequila.

Mezcal production is one of the most painstaking and patient pursuits in the world of beverage. Many maguey – the most common agave for mezcal production – grow wild. The security and consistency

VIGNETTE

of cultivation does not exist. These maguey can take over ten years to mature and can only be used once. There is no "next year's crop." One plant grows its whole life for one moment. And although the agave hearts are giant, only about four to five bottles of mezcal are made from each one.

Mezcal made me feel small. And it made me want to be more patient. I sipped mezcal on the beach and realized that it was okay to set limitations for myself. I didn't need to work the most, party the most, or achieve the most. I could take time to sleep, rest, eat well, and exercise. It didn't mean I was giving up. I could accept what was and pursue something better at the same time. I could sip a little mezcal to feed my enchantment with the spirit without needing to line up shots of it.

229

Mezcal and Friends Map

Mezcal is defined as a spirit distilled from agave. Over 30 different types of agave are permitted, as well as production in towns across nine different states.

Tequila is required to be made from at least 51% blue agave and can be produced in five different Mexican states. Practically, the difference is the method of production: most mezcal is cooked in underground pit ovens, giving the spirit its overt smokiness, while tequila is cooked in above-ground brick ovens or autoclaves that do not impart flavor.

Bacanora is a spirit made from one type of agave (*Agave angustifolia*) and in one state (Sonora). It is made traditionally like mezcal but tends to be a bit less smoky and leaner (which producers believe is due to the cooler climate of Sonora).

Sotol is a spirit that is produced in three Mexican states (one of which overlaps with mezcal production). Unlike the former three spirits, sotol is based on a succulent called *Dasylirion wheeleri* rather than agave. It is traditionally smoked like mezcal and finds its distinction in a striking earthiness created by the dry terroir of the Chihuahuan desert.

Sonora

Chihuahua

Coahuila

Durango

Zacatecas

Tamaulipas

San Luis Potosí

Nayarit

Guanajuato

Jalisco

Michoácan

Puebla

Guerrero

Oaxaca

Vodka

Beth and I had always referred to Ed as our godfather.

We didn't really know what this meant, and he wasn't in any religious sense an actual godfather, but we knew that he and Rob were the ones who would take care of us if anything happened to our parents. Obviously, we didn't want anything to happen to our parents, *but it would be cool to go live with Ed and Rob*, we thought.

Ed and Rob had gone to law school with my mom, and their circle of friends stayed close. We would get together with their families every Memorial Day weekend in La Quinta. We grew up with this second family. Beth and I were the oldest kids of the group, so we felt like adults when the younger kids came along. We developed many close relationships with my parents' friends, and Ed and Rob were the closest. Rob was the silent and stoic type, the straight man (even as a gay man) to Ed's comic lead. Ed was always joking around. He always had a gentle lilt and lift in his voice that said *nothing is ever that serious.*

At night on those weekends, when we convened, our cheeks rosy and skin sticky from a day of playing by the pool, Ed used to play a game with all the kids. He would have a handful of cash and change in his pocket, and if anyone guessed the exact amount, they would win the money.

"$1.42?"

"$4.88?"

"$9.17?"

The guesses went around. Ed had a knowing grin on his face the whole time. He took in every guess and pondered it for a moment, like it could be the right one. When it got around the circle, Ed pulled the wad of money out and began to count.

"$122.45. Did anyone guess that?"

He knew that no one had guessed that. No one's guess had surpassed $20, and *$120* was an amount inconceivable to our young minds. Our jaws hung slack as we looked around the circle, each pondering what it would have been like to be a *millionaire*.

Ed tucked the money back into his pocket. "Better luck next time," he said with a smile.

VIGNETTE

—

My first inkling that something was wrong was in New York. I was spending a summer there while in college, and I met Ed and Rob for a drink one night while they were visiting the city. Ed ordered a Maker's Mark Manhattan, up, with three cherries. I was at the infancy of my drinking career and probably ordered a mojito, or maybe sparkling water. Rob nursed a vodka martini.

We had barely been talking five minutes and Ed was done with his drink. He called the waitress for another. It was also gone in five minutes, and then there was another. This was repeated I'm not sure how many times. Rob and I each had one drink. I remember Rob seeming somewhat uncomfortable, but we never talked about what had happened that night. I'm not sure if I talked about it with anyone. At the time, I just thought Ed had a high tolerance and really enjoyed Manhattans.

It wasn't until several years later that my mom told me Ed had alcoholism. Now that I was in the infancy of my career in booze, I tried to reconcile that what I did for a living was so harmful and destructive to some.

Rob, my mom, and several other close friends in Ed's life dug in deep: there were interventions, there was rehab, there were long talks, there was lots of crying. Everyone was fighting hard for Ed, but Ed didn't seem able to fight hard for himself. The pain was too deep, and the only thing that would relieve it was alcohol. Bourbon Manhattans turned into a water bottle full of vodka.

At this point, I had several jobs and obligations, and my trips to California for Memorial Day became less and less frequent. I didn't get to witness Ed's deterioration, except for one lunch I remember in the later stages of his disease. The same Ed was still in there, fighting to come out, but there was a dullness to his eyes and a distance to his soul. The lilt in his voice was gone, and suddenly everything *was that serious*.

I couldn't imagine how Rob coped. He had tried everything to help Ed. At one point, he moved out of the house, reasoning that if Ed could see that he was losing the person he loved most in the world because of alcohol he would quit. But leaving Ed didn't make Ed – or himself – better, so Rob chose to stick by his love's side, even if it meant watching him die.

233

VODKA

I was working at Eleven Madison Park when I got the news from my mom that Ed had passed. I cried in the cellar. Until that point, I had always thought there was a way it could turn around. That he could get better. It seemed unbearably tragic, unfair even.

The ultimate injustice, though, is how the world views alcoholism. I get it: it's an incredibly odd and cruel disease, one that relies on a person picking up a bottle and bringing it to their lips in order to do its damage. Because of this, like most cases of addiction, it is viewed as a weakness and a character flaw rather than a disease. Words like *drinking problem*, *drunk*, and *lush* at once blame the victim and minimize what they are going through. Ed lost his battle with alcoholism, just like someone can lose a battle with cancer. The fact that Ed's defeat depended on his own actions should not make him less worthy of our sympathy. Perhaps it should even make us feel more. Not only did alcoholism take Ed's life, it crippled his judgement and strangled his freedom of will.

Ed was a private person and sensitive about how people perceived his drinking. But I know he would support his story being told with this intention: that it might motivate others to seek help and for our society to more broadly appreciate the challenges of alcoholism and addiction.

Resources for Alcoholism

Many people in this industry, and in the world, have "drinking problems." This can be seen as drinking too much in social settings, relying on alcohol to feel relaxed, or turning to alcohol to deal with distress. A lot of us have been guilty of this at one time or another in our lives, and these behaviors can warrant seeking help. But alcoholism is a different issue. It is a chronic disease. It is a predisposition to be addicted to alcohol, to need it every day, and to not be able to stop drinking. For people who suffer from it, help is necessary.

Alcoholics Anonymous

Now an international fellowship, AA's stated purpose is to enable its members to "stay sober and help other alcoholics achieve sobriety."

Drinkaware

An independent UK alcohol education charity working to reduce alcohol-related harm.

Drink Wise

An organization committed to creating a healthier drinking culture in Australia.

EU Health Promotion and Disease Prevention

Information on alcoholism, alcohol abuse, and their health and economic burden.

National Institute on Alcohol Abuse and Alcoholism

A US institute that conducts and funds research on the causes, prevention, and treatment of alcoholism.

Reach Out

An online mental health organization, providing support, tools, and tips for young Australians and their parents.

Substance Abuse and Mental Health Services Administration

A branch of the US Department of Health and Human Services whose mission is to "reduce the impact of substance abuse and mental illness on America's communities."

World Health Organization

The WHO offers resources and research on preventing and controlling alcoholism and alcohol-related harm.

F.E. TRIMBACH *CLOS STE. HUNE RIESLING* Alsace, France
DOMAINE WEINBACH *PINOT GRIS VENDANGES TARDIVES* Alsace, France
DOMAINE ZIND-HUMBRECHT *GEWÜRZTRAMINER* Hengst Grand Cru, Alsace, France

236 Alsatian Wine

When Jon and I became a couple – for real and finally – it was under the assumption that this was it.

We both knew who the other was, we had been through the ups and downs, and there was an inextricable connection between us. Marriage was on the table from day one, and neither of us was scared by this.

We waited a little while to get engaged, though. We wanted to do some things the old-fashioned way. Jon asked my parents for permission. My mom said yes and my dad cried. Jon picked out and bought me a vintage engagement ring. Then one night in January, when I was in bed early and he had just gotten home from work, he knelt beside me and opened the box. I wish I remembered more clearly what he said. Something about loving me and wanting to spend the rest of his life with me, I'm sure. I was distracted by the ring, the moment, and I blurted out a teary "yes!" before he could even finish talking.

The next day we were meeting about 15 friends for brunch (how wine people do brunch). This was planned independently from the engagement – merely a special wine restaurant we'd been meaning to go to. We decided to let our friends discover the ring rather than make an announcement.

The restaurant was a couple of hours outside New York, so that morning we all piled into a giant rental van to make the journey

VIGNETTE

up together. The car ride went by without anyone noticing the ring; perhaps I wasn't doing enough hand-gesturing. When we settled in for brunch, we requested as many wine lists as the restaurant could budget us. I was afraid to see what would happen as a room full of sommeliers tried to make wine decisions together. Not enough triage, certainly, as within a few minutes six or seven bottles of wine were on the table. I ordered a bottle of Alsatian Riesling. It was a humble bottling, with mid-term age on it, from a great producer. It reminded me of my engagement ring. Traditional and simple, it shone because it wasn't trying too hard to be flashy. There was a subtle and effortless opulence to it. We all agreed, it was one of the best bottles of the day.

By mid-meal, with no one noticing the ring, Jon – slightly tipsy by now – began to get agitated. "What's wrong with these people?" he whispered. "Aren't they supposed to be our friends!?" I laughed at Jon's tendency for hyperbole. I told him to be patient. That it didn't have to be the focus of the day.

We were standing up to get some air with a few friends when Jon's patience ran out. He grabbed my left hand by the ring finger and wagged it in their faces. "Notice anything?" he enquired. I laughed sheepishly, pulling my hand away, but it was too late. Soon, everyone in the room knew, and impromptu speeches broke out. It was slightly embarrassing, but I also realized that it was important to celebrate the moment with our friends. Jon had been forward, brazen, and somewhat obnoxious, but ultimately, he had been right. We drank wine until the early evening, then napped on each other's shoulders on our drive back to the city.

ALSATIAN WINE

Pairing the Noble Grapes of Alsace

The cuisine and wines of Alsace have evolved together over time, both taking on distinct flavors that pair naturally with one another. The below are common pairings for the four noble grapes of Alsace, with dishes and foods that are some of the most iconic in the region.

The orange blossom and white plum of dry Muscat enhances and mellows the creamy vegetal note of white asparagus.
Try: Dirler-Cade *Muscat*, Grand Cru Saering

The spiciness, sweetness, and bitterness of a classic Grand Cru Gewürztraminer cuts through the fat and pungency of Munster cheese.
Try: Domaine Zind Humbrecht *Gewürztraminer*, Grand Cru Hengst

Vendanges Tardives (late harvest) Pinot Gris, with its mushroomy sweetness, creamy texture, and slight bitterness, lifts and cleanses the richness of foie gras.
Try: Domaine Weinbach *Pinot Gris Vendanges Tardives*

The classic Alsatian tarte flambée – a crusty flatbread with fromage blanc, onions, and lardons – demands the acidity and diesel smokiness provided by Riesling.
Try: F.E. Trimbach *Clos Ste. Hune Riesling*

GIACONDA *ESTATE CHARDONNAY* Beechworth, Victoria, Australia
BASS PHILLIP *PREMIUM PINOT NOIR* Gippsland, Victoria, Australia
MOUNT MARY VINEYARD *QUINTET* Yarra Valley, Victoria, Australia

Aussie Wine

I never would have thought
I'd end up in Australia.

But by November 2016, I had been at Eleven Madison Park for two years – two years of hard work and astounding growth – and it was time for something different. My original plan was to study for the Masters and work on a yet-to-be-named writing project, but soon after I gave my notice at EMP, I was offered an opportunity to work at one of the best restaurants in Australia. It came up casually at first, and both Jon and I laughed and brushed it off. *Move to Australia? That's crazy.* But something about it made me want to take the interview anyway.

I had a Skype call with Ben Shewry, the chef-owner of Attica. We talked for a few hours. It was nothing in particular he said, but just a feeling that this was right. Like most major decisions in my life, I made this one on instinct. And Jon did too. Jon had never before made a life decision based on anything except his career. But he came home from work late that night, sat down on the couch and, without any conversation, said, "Let's move to Australia."

—

In New York, we drank European wine. Our home cellar (that is, wine fridge, under-the-couch shoeboxes, pupitre in the kitchen, wall-mounted shelves, and stand-up wine rack) was filled with Chablis, Chambolle-Musigny, Barolo, Côte Rôtie, and the Rheingau. We loved the humble and grand Sangioveses of Chianti Classico, the spritzy reds of the Jura, and the up-and-coming producers of Germany playing outside of traditional Prädikat levels. A few bottles from the New World dotted our shelves: often wines for blind-tastings (Mendoza Malbec, Central Coast Viognier), a few older vintages of wines we were curious about (Littorai and Calera Pinot Noir), and a rare bottle or two of American wine we bought because we liked drinking it as much as European wine (Arnot Roberts Trousseau and Hermann J. Wiemer Riesling come to mind).

I had been on a few wine trips in Northern California and visited the Finger Lakes once, but American wine travel wasn't a huge priority for me. To this day, I've never been to the vineyards of the Willamette Valley, Santa Barbara, anywhere in Washington State; and – probably the most embarrassing of all – I've never visited the wine regions of Long Island. I chose to stay home and drink European wine rather than venture out into my own backyard.

I was excited to explore Australia's wine regions, but did not anticipate actually liking the wines. I imagined visiting Beechworth, Yarra Valley, and Gippsland, then returning home and enjoying a bottle of Barolo.

My first visit to an Australian bottle-shop was a rude awakening. I walked down the aisles of my familiar favorites and did a double take. Bourgogne Rouge starting at $75, Chianti Classico no cheaper than $50, current-vintage German Spätlese starting at $60 – and forget about buying a decent bottle of bourbon! If I wanted a $30 bottle of retail wine, it was going to be either bottom-of-the-barrel European (bottles I knew would cost less than 5€ over there) or Australian.

I left the wine store that day with bottles of Yarra Valley Nebbiolo, Adelaide Hills Pinot Meunier, and Strathbogie Ranges Riesling. I spent less than $100 on the three, but I was nervous about what I'd find in the bottles, and apprehensive about my wine-drinking future in Australia.

To my great surprise and satisfaction, all three were superb. The Nebbiolo was fragrant and lifted, graced by just enough **VA**, with the roundness of juicy red fruit upfront balanced by a drying finish. The Pinot Meunier was slightly hazy, garneted, and transparent, with a supple and velvety texture. And the Riesling deviated from the Australian norm of stinging acid and zero sugar, with about 10 grams per liter **residual sugar** and singing aromatics.

The first weekend we got the chance, Jon and I rented a car and drove out of Melbourne. It was Easter weekend, celebrated like Americans celebrate Labor Day, as the last weekend of the summer. There was a big festival in Beechworth, so we decided to drive there. This was Jon's first time driving on the left side of the road, and after several near-scrapes and harrowing freeway on-ramps, we were on our way.

We had fallen into normal road-trip activities when Jon suddenly squealed with delight at a freeway sign: "It's Nagambie Lakes!" I lit up, too. Nagambie Lakes was a place we had studied but never thought we'd actually see – like Narnia or Hogwarts. We scurried across three lanes to make the exit, and as we drove into Tahbilk Winery, we felt a weight of historical significance.

The cellar door, as Australians call a tasting room, was like an old Gold Rush saloon. We learned that Tahbilk is the largest landowner of Marsanne vines in the world (they have more Marsanne than is planted in all of Hermitage), tasted the wines, and ate lunch overlooking one of the lakes.

Back at work, my sommeliers were nonplussed by my weekend adventures. One of them, who tends toward the more "natural" styles of wine, could not believe I'd had lunch at Tahbilk and enjoyed sparkling Shiraz (what Australian winemakers have typically called their Syrah). We bantered playfully about my pull toward "industrial wine," and I countered that he enjoys the stench of bacterial overgrowth. But when it came down to it, what he couldn't believe was me taking so much pleasure in what he considered the simple and pedestrian wines of his home country. To me, they were exotic and new, but to him, they were old hat. I thought about the wines of my previous backyard and how much better my new Australian backyard was.

The weekend after Jon and I got back from Beechworth, we had dinner plans with some newfound friends. We decided to do BYO Szechuan last minute and ducked into a dodgy bottle-shop to grab a few wines. We ended up with (relatively cheap) Tasmanian Pinot Gris, McLaren Vale Grenache, and sparkling Shiraz – worth a shot, we thought. The Grenache ended up being a home run with the bold and spicy food. At 14.5% alcohol, it stood up to the flavors, with a kiss of residual sugar that licked some of the heat off the palate.

I started to go on about how impressed we were with the wines of Australia and that American wines weren't nearly as exciting.

Jon interjected. "There's a lot of really great American wine."

"Yeah," I continued, perhaps a little too much wine under my belt, "but not like what we've had here. I mean, the diversity, the character, the value. It's so much better!"

Jon gave me a look that said he wasn't done with this conversation, but dropped it in the face of our new acquaintances. On the way home, he picked it up again. Hard. "I don't know why you're hating on American wine so much."

"I don't know why you're defending it so much," I replied. "It's not like we ever drank a lot of American wine. And we've been excited to drink Australian wine here."

"Yeah, that's only because we can't afford Europe over here."

"That's not true."

"Maybe you never gave American wine a chance!" Jon yelled, storming up to our apartment. I had no idea why we were fighting over something so stupid. We went to bed, backs turned to one another, no more words spoken.

The next day, Jon apologized. I did, too. It struck me that Jon missed America. Perhaps not the wines, but our friends, colleagues, and family. My blind dismissal of American wine hit a nostalgic chord, and he lashed out. But I also realized that I never really did give American wine a chance. I had studied and tasted the wines of my home country for exams, but I never looked at them through a lens of excitement and curiosity. It took moving to Australia, and seeing someone else's backyard through that lens, to show me I'd never afforded America the same chance.

Kangagories

I started a short-lived tradition at Eleven Madison Park of playing a certain categories game at pre-service line-up once a week. I came up with custom categories, specifically about the food, beverage, and hospitality industries. When I went to Attica, I took the tradition with me. It is a fun team activity that can be played quickly and without a lot of materials. And it rewards people with genuine curiosity and knowledge about our industry. Choose a starting letter, set a time limit, and fill in as many as you can. All answers need to be Australia-specific.

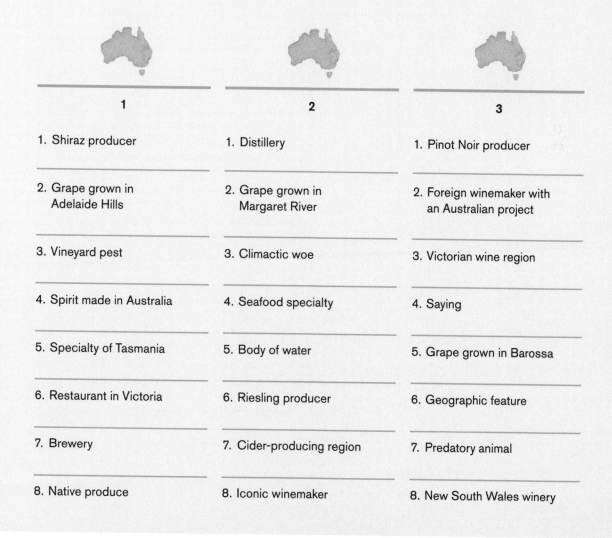

1

1. Shiraz producer

2. Grape grown in Adelaide Hills

3. Vineyard pest

4. Spirit made in Australia

5. Specialty of Tasmania

6. Restaurant in Victoria

7. Brewery

8. Native produce

2

1. Distillery

2. Grape grown in Margaret River

3. Climactic woe

4. Seafood specialty

5. Body of water

6. Riesling producer

7. Cider-producing region

8. Iconic winemaker

3

1. Pinot Noir producer

2. Foreign winemaker with an Australian project

3. Victorian wine region

4. Saying

5. Grape grown in Barossa

6. Geographic feature

7. Predatory animal

8. New South Wales winery

244

Natural Wine

"Natural wine" has, unfortunately, become a contentious topic.

In certain cities – Copenhagen and Sydney come immediately to mind – you must come down firmly on one side or the other. You're either a "natural wine" restaurant/bottle-shop/importer/distributor, or you aren't. This strict divide, this either/or, is what I find the most harmful. The antithesis of an inclusive and warm wine society.

I want to be precise in this discussion, so I am using the definition of "natural wine" provided by RAW WINE™, an international wine fair devoted to "low-intervention, organic, biodynamic, and natural wine." The fair is run by a French **MW** living in London; the approach is educated and reasoned.

> Natural wine is farmed **organically (biodynamically, using permaculture** or the like) and made (or rather transformed) without adding or removing anything in the cellar. No additives or processing aids are used, and 'intervention' in the naturally occurring fermentation process is kept to a minimum. As such neither fining nor (tight) filtration are used. The result is a living wine – wholesome and full of naturally occurring microbiology.

This all sounds well and good, but there are some problems.

Problem #1
The name "natural wine" implies there is "unnatural wine." But where is the line drawn?

The co-opting of the term "natural" to apply to this genre of wine was one of the greatest marketing moves of all time. Who wouldn't want their wine to be *natural*? And does this mean the other stuff is *unnatural*? Where is the line drawn?

Those who've created this divide have the burden of defining it. And that's where the problem starts. Wine production requires some level of intervention. But how much intervention is too much? Hand-harvesting is favored. Inoculated yeasts are off-limits (but Champagne gets a pass). Fining and filtration should be kept to a minimum. **Spinning cones** are a no-no to concentrate must, but **saignée** is probably okay. No **chaptalization** (though Burgundy will still find a home). No acidification (but sherry is fine). And – most contentious of all – all additives are strictly off limits.

Except – sometimes, in minute amounts, as sparingly as possible – the use of sulfur. Sulfur is a natural biproduct of wine production, and small amounts exist in wine without any additions. Most wine – I couldn't find a statistic, but I would imagine upwards of 95% (and possibly over 99%) – also receives sulfur dioxide additions at one or more points in the winemaking process. SO_2 acts in an antioxidant, antibacterial, and disinfectant capacity. Most winemakers find it invaluable in making pure, clean, and unflawed wine.

Sulfur additions have, quite unfortunately, become the lynchpin of the "natural wine" discussion. Some staunch supporters have a zero-tolerance policy towards sulfur. RAW has a more moderate approach and calls for it to be used sparingly, never showing an excess of 70 grams per liter in a finished wine. The need to define an upper limit makes sense in the context of accepting submissions for a fair, but it feels undeniably arbitrary. Why should a wine that sits at 69 grams per liter be acceptable, while the wine that has 71 is considered unacceptable?

Wine is, by definition, unnatural, manipulated, and produced. The extent of this "intervention" has now become the issue. It's a spectrum with many shades of gray and can't be reduced to a strict binary equation, with benevolent on one side and malevolent on the other.

Problem #2
Not all "natural wine" is good wine.

As the "natural wine" movement has grown and the line between *us* and *them* drawn more firmly, there hasn't been a huge distinction between quality and lack thereof in this arena. Many "natural wine" producers are extremely diligent. They grow or source impeccable fruit, they have the highest standards of hygiene, they know how to

245

NATURAL WINE

handle fruit that will not see much sulfur, and they are responsive to the needs of the vintage. It is not the easiest way to make wine. In fact, it's probably the hardest. There's no safety net of chaptalization, acidification, sulfur additions, or yeast inoculation. The fruit and its handling must be pristine. These winemakers do not follow the tenets of "natural winemaking" because it is a trend or a marketing tool; they do it because they know they can make great wine this way.

Not every producer is so proficient or so scrupulous (or even has the right prerequisite conditions). Many wineries have thrown their hat in the "natural" ring – for whatever reason – without knowing the difficulties involved. These wines turn out well and truly flawed: overrun with bacterial overgrowth that masquerades as "wholesome microbiology" within these circles. And everyone is too afraid to criticize these wines lest the whole house falls down.

Even RAW WINE™ sets forth a charter of quality that has nothing to do with quality. It is a list of criteria that a wine must meet, including limited use of sulfur, no yeast inoculation (except for secondary fermentation), all organic/biodynamically grown grapes, only hand-harvesting, no additives, no "heavy manipulation," limiting fining/filtration, etc.

But nowhere does it discuss how these wines taste. Are they free of distracting flaws? Are they delicious? Do you want to drink them? It's as if meeting the criteria is enough to get a seat at the table, never mind what is actually in the bottle.

Problem #3

What is meant to be a pure, "natural" expression of the land is often obscured by flaws.

Picture a drinking well, clouded with naturally occurring solids and debris. Does cleaning this well by running the water through a filtration system result in less purity and clarity? Or perhaps it gives it more?

Picture a face, cleared of a rash through the use of antibiotics. Do we get a clearer picture of the person, or have the antibiotics obstructed nature?

Picture a body lotion, enhanced with a naturally occurring antioxidant so it doesn't turn brown in the bottle. Is this product marred by the addition, or does it gain integrity?

Picture yogurt, using cultured lactic bacteria instead of letting the milk ferment on its own. Instead of mold and off-flavors, we get a clean, delicious, and consistent product.

The analogies go on. We accept all sorts of treatment, manipulation, and additions to preserve the clarity and integrity of products in our daily lives. My argument here is not whether or not these additions and interventions are "natural," but whether they preserve purity of expression. Much "naturally" made wine tastes like volatility, oxidation, **Brettanomyces**, **reduction** (caused by excess solids in the bottle), and bacterial overgrowth – obscuring the expression of a grape or a place. I don't consider any of these "flaws" absolutely unacceptable (except maybe bacterial overgrowth). The other four, when kept in check, can enhance the character of a wine. But when these things run rampant, they usurp any sort of terroir expression. What was supposed to be a pure and natural expression of a place becomes an expression of winemaking (or lack thereof): the exact antithesis of the purported goal.

Problem #4
The discussion has distracted sommeliers from hospitality.

As a sommelier, I believe my first duty is to my employer. If I can't create a financially sustainable beverage program, there is no business. There are no customers. There are no wine purchases. This obligation must come first.

My second duty is to my guests. If the guests aren't happy, there is no business either. My duty to my guests is to give them a great experience, which means different things to different people. For some guests, this means turning them on to a new style of wine or region. For other guests, it means helping them choose the bottle of the exact style they already like. For some, it means providing engagement, education, and information. For others, it means leaving them alone and letting them enjoy their own company. My needs – my ego, my preferences, my agenda – do not come into play.

My third duty is to the winemakers I represent. They are purveyors for the restaurant and, as such, are always treated with respect. I respond to every email. I try to take every appointment that is requested of me. I never haggle; I always pay the asked price for a product if I think it's worth it. I want to give space to winemakers who are creating beautiful products that make people happy. But I don't believe bringing their story, their philosophy, and their agenda to guests is my primary responsibility. If guests want this information, I am happy to provide it, but I don't force it on anyone.

A restaurant is not a place to be lectured, to be made to feel stupid for your preferences, or to be challenged to be adventurous (if that's not what you want). It is a place to be offered warm, welcoming, and genuine hospitality – whatever that means for each guest.

—

In conclusion, I want to make it clear that the problem is not "natural wine." The ambitions of low intervention and purity of expression are indeed noble. And they are ambitions I agree with wholeheartedly. The problems arise when this conversation becomes exclusive instead of inclusive, when it encourages sloppy winemaking, when it obscures rather than enhances expression, and when it takes precedence over hospitality.

I propose we drop "natural wine" and get back to talking about "great wine" – great wine made by people who are good to the world and good to each other, and served by people who believe in hospitality.

VIGNETTE

Somm Survey: How do you define "natural wine" and what is your opinion on the topic?

"The term 'vin natur' dates back to early the 20th century when farmers in Roussillon were protesting negociants' addition of sugar to their musts. In the '80s, the term resurfaced and took on a different usage: organic viticulture, additive-free winemaking, etc. Today the term literally means nothing as it has been bastardized and misused."
Caleb Ganzer

"Natural wine is like giving someone a heads up that what they are about to drink is going to be more expensive than they would expect, while having the added bonus of potentially containing faults that could have been avoided."
Benjamin Hasko

"I think anytime someone sees anything as dogmatic, you stop pushing creativity and experimentation."
Cedric Nicaise

"It's the 'flat Earth movement' for winemaking. I'm all about being a non-interventionist, but to what point? If you want a science experiment, drink Sherry or Champagne."
Eric S. Crane

"I define it as a category of wine with a misleading and bad name. All winemaking has some sort of intervention, or it would be vinegar. I'm a generalist and like delicious wines of many styles, from all over the world, ideally made by cool people."
Jill Zimorski

"I'm happy the natural wine movement is generating a new level of transparency and consumer interest. It's important for the wine world to formally define and regulate this term, though, as it's become a dogmatic marketing phrase that has opened the door for mediocre producers to group themselves with true masters of the craft."
Jordan Salcito

"Some of the finest wines in the world are made this way, both today and historically. And while it's true that sometimes this results in flawed and/or unpalatable wines, bad wines are also made using more conventional or modern methods, perhaps more often."
Kelli White

"My personal opinion is that everyone's perspective is valid; everyone's taste is correct; can't we all just get along?"
Stevie Stacionis

"If the wine performs in the glass, I don't care what the winemaker's philosophy is."
Mia Van de Water

"This is the greatest heist in wine history. The one good part that has come from it is the frank discussion and earnest attempt on the part of many producers to do as little as possible to the wine and to deliver something that pristinely speaks of its place as opposed to just its process."
Richard Rza Betts

NATURAL WINE

South American Wine

250

I had intended to take the MS exam again in 2017.

I'd started studying the summer before and was building some momentum, but I knew I couldn't move to Australia, devote the time required to become proficient at Attica, and still pass theory. So I waited one more year.

I paved out a year-long study plan. I created my own study guides for each region. I traced maps from the internet and added important producers, geographic features, and vineyards. It was time-consuming, but it exponentially enhanced my understanding. I used various resources to source information on history, climate, soil type, wine styles, and current trends – then wrote it all out in my own words. I created my own producer profiles and vintage reports.

I charted any pertinent wine laws: permitted grapes, methods of production, aging requirements, **residual sugar** allotments, minimum must weights, etc. I went all in. It was the only way I knew how to do it.

My favorite thing about studying and tasting for exams is finding points of interest in regions I have previously found uninspiring or intimidating. Regarding the latter, I began delving into the flavors of the Loire that I had avoided since TopSomm in 2015. In regards to the former, South America had always been one such region for me, never quite hooking me in.

When I went deep into South America this time around, I discovered many wines I grew excited about. One-hundred-year-old País vines in the southern valleys of Chile, "**pipeños**" of traditional Chilean varieties, Chasselas from Itata, old-vine Carignan in Maule, Sémillon in Patagonia, Brazilian Teroldego, and Tannat from Uruguay made like **Barolo Chinato** are just a few of the compelling things going on in South America. These hooks allowed me to gain traction in my studies.

But when it came to blind-tasting South America, I struggled to find my hook. I found the wines extreme. Torrontes was floral and piney. Carménère smelled like charred shishito peppers. And Malbec, so extremely…boring. I had never understood the appeal. The fruit was electric purple, confected on the nose, but green on the palate. Sometimes there's an oak presence, but the average Malbec is not expensively made (i.e. not a lot of new oak). There were subtle notes of purple flowers, bitter coffee, and tobacco leaf, but overall its characteristics were not as pronounced as many other grapes.

And, perhaps adding to my disdain, I always missed Malbec in blind-tastings. I had been known to call a Malbec as Northern Rhône Syrah, California Cabernet, Chilean Cabernet, Australian Shiraz, Zinfandel, and even American Pinot Noir: pretty much anything but Malbec.

"It's because you don't understand it," Jon explained.

"Sure I do," I retorted, and went into my explanation of Malbec.

"No," he cut me off, much to my aggravation. "Argentine Malbec is what happens when you make wine in a warm climate, but at extremely high altitude. The winds are fierce and the conditions change drastically from day to night. There is bright, jammed fruit with a vein of greenness. There is structure that gives way to softness. Oak is welcomed, but not necessary. It's a study in contradictions."

Jon proceeded to slip Malbec into flights for me in the months to come. Malbec next to Zinfandel. Malbec next to Chilean Cabernet.

Malbec next to Saint-Joseph. The turning point came when I approached a wine that I initially wanted to call Napa Cabernet. It was ambitiously made, had ripeness and structure, and a fair bit of oak influence. But something was different. The fruit character was bluer and more purple than it was black, more berry than it was currant. The **tannins** were less regal and austere; they were plush and easy. The palate tightened up and got less ripe, providing a burst of refreshing acidity. *This*, I thought, *is what Malbec is all about.*

I called the wine correctly. I grew to understand Malbec better that day. And, perhaps most important of all, I understood why people drank it. Its appeal finally made sense.

Jon cooked steaks on the grill that night and we drank the rest of the bottle.

"I think this is the first time I've ever *drunk* a Malbec," I remarked.

"I know," Jon said. "And that's why you didn't understand it."

I hated – so much – when he was right.

Altitude Chart of South America

The landscape of South America is defined by dramatic spreads of elevation: from sea level to almost 7000 meters above it. The wine regions are similarly influenced by elevation, with high altitudes providing diurnal swings and exposure to the elements. Taking a look at the literal highs and lows is a valuable exercise.

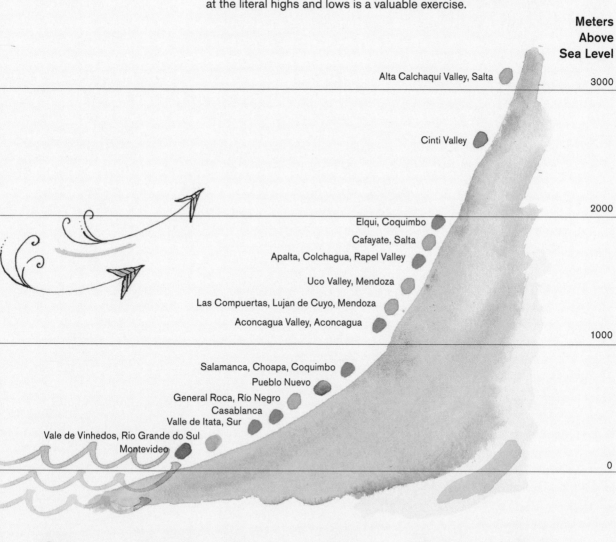

Meters Above Sea Level

Alta Calchaquí Valley, Salta — 3000

Cinti Valley

— 2000

Elqui, Coquimbo
Cafayate, Salta
Apalta, Colchagua, Rapel Valley
Uco Valley, Mendoza
Las Compuertas, Lujan de Cuyo, Mendoza
Aconcagua Valley, Aconcagua

— 1000

Salamanca, Choapa, Coquimbo
Pueblo Nuevo
General Roca, Río Negro
Casablanca
Valle de Itata, Sur
Vale de Vinhedos, Rio Grande do Sul
Montevideo

— 0

Bolivia   Argentina   Chile   Brazil   Uruguay   Peru

DOMAINE DU VIEUX TÉLÉGRAPHE *LA CRAU* Châteauneuf-du-Pape, Rhône Valley, France
YALUMBA *VINE VALE GRENACHE* Barossa Valley, South Australia, Australia
ALVARO PALACIOS *L'ERMITA* Priorat, Catalonia, Spain

Grenache

A few months after I began studying
again, Jon returned to the US to take
his MS tasting exam.

He had passed theory and practical in 2017. Jon had been tasting
somewhat erratically. He would have miraculous tastings – where
he nailed 15-year-old Hermitage Blanc, parceled the difference
between young Carménère and Chilean Cabernet, and gave the
exact vintage and commune of mature Bordeaux. He also always
nailed Châteauneuf-du-Pape, in all its multitude of forms: floral and
fruity, funky and **bretty**, mature and dry, purple and massive.

But he also had tastings where he missed the obvious:
Chablis, Chianti Classico, Saint-Joseph, and Aussie Grenache
eluded him at times. I understood the emotions that attended this
and observed the ups and downs with each flight. If Jon tasted
well, he felt confident, happy, optimistic. If he tasted poorly, he
was grumpy, sullen, and self-loathing.

Jon's final flight, the day before he left, was weak. He cursed,
probably threw things, and stormed out of the room.

"I'm not going to pass," he said that night as we lay in bed.

"Not with that attitude," I replied.

He glared at me.

"Don't overcomplicate it. Don't overthink it. But you *have* to
believe that you deserve to pass."

"Yeah, I guess," he said. He turned his back to me and fell asleep. He kissed me on the forehead the next morning as he left the house at dawn.

Results were set to be given at 10am his time in St. Louis. This was 3am my time. I tried to stay up, but drifted in and out of sleep. I finally woke at around 3:30am to a phone call from Jon.

"Baby, I passed! I passed…!" Jon sobbed. It was the first time I had ever heard him cry. I cried too. He was a Master Sommelier. And he finally had the pin to prove it.

GRENACHE

Grenache Word Search

Words can go vertical, horizontal, and diagonal, as well as backwards and forwards. No spaces, no accents. Happy hunting!

256

```
E Y Y O V U A P T T Y P V W T Y V P U P
R E H P A R G E L E T X U E I V I S A A
D T A I E W C D E N V N N T L E C A E S
E A T G J N J H I T A F E C K A A D N O
V R E N V D O D W D T R N E Z Q N N N R
R O A A Y Y R H R D R E N L G N N O O O
U I U N N A X A R E L I R V E B O G B B
O R N B C Y C C T U E V C I S V N I I L
M P E S P I I N O R D U A I A B A G R E
K N U B P N O B S J J S G C R L U T N S
U M F Q S I R K D J H A E I C I C R E Q
F Z D A R U L L B A O X M T R A L F H G
W F U R O O B A N Y U L S E O R R L J A
P L P B O P I Q U E P O U L S C A E O L
T S A F M C L A R E N V A L E B H G S L
K B P R A S T E A U H A R Y S X E R I E
L M E S A R Y E U Q C A V H L N A R H T
X N R B N L A T I M R E L C O Y X U R S
U K R O U S S A N N E E O X A L J I C Y
E S I O N U O C U T B I Z S W C G H U B
```

Banyuls
Barossa Valley
Bourboulenc
Cannonau
Châteauneuf-du-Pape
Cinsault
Cirillo
Clairette
Côtes du Rhône
Counoise
Gallets
Garrigue
Gigondas
Henri Bonneau
James Berry
L'Ermita
McLaren Vale
Mourvèdre
Muscardin
Paso Robles
Picardan
Piekenierskloof
Pignan
Piquepoul
Priorat
Rasteau
Rayas
Roussanne
Syrah
Tavel
Terret Noir
Vaccarèse
Vacqueyras
Vieux Télégraphe

VIGNETTE

Women's Wine

I'm often asked what it's like to be
a woman in the wine industry.

Sommeliers have traditionally been men, and to a certain extent the top positions in the field are still dominated by men. As of going to press, there are 256 Master Sommeliers in the world. Thirty-one of them are women. If you look around the world at the head sommelier and wine director positions in top restaurants, they are mainly filled by men.

I imagine most interviewers are hoping for stories about chefs grabbing my ass and guests refusing to acknowledge my expertise. This has not been my experience. But here's what I do have to say: being a woman in the wine industry is not the tough part. Simply being a woman is what's tough.

Let's start with appearance. For whatever reason, it is open season for people (of either gender) to comment on a woman's appearance in a way that would never be directed at a man. During the airing of *Uncorked*, a prominent American newspaper ran a piece about the show. The journalist (a woman) wrote of me: "30-year-old Jane Lopes is the pretty but perpetually nervous sommelier at Eleven Madison Park." I was less offended by her characterization of me as a nervous novice than by her need to comment on my appearance. The implication was that being pretty was a consolation prize for not being good at my job.

WOMEN'S WINE

I've been told by a male boss that I have pretty toes and I should paint them. I've been told that my hair is too curly, and that my hair-tie is the wrong color. Everyone comments on my height. I've heard female co-workers being told they wear too much make-up, or not enough. And I have not worked in the worst situations, by any means. I've heard horror stories from friends about being harassed, sexualized, insulted, and demeaned based on their appearance.

Generally speaking, most women don't have the luxury of not thinking about their appearance. And for most of us, it is not a simple, or cheap, undertaking.

Hair management costs a considerable amount of time and money for many women. Unlike a man's daily shaves and monthly visits to the barbershop, a woman's regimen of maintenance, styling and removal is a complex matrix of daily, weekly, and monthly management that spans every corner of the body.

Then there's skin care. I had severe acne in my teens and early twenties, which only let up after two rounds of Accutane. Every morning I was nearly brought to tears as I carefully painted concealer on my face to hide these blemishes. I watched as boys in my class also developed acne, yet never felt the need to cover it up.

The acne subsided in late college but I still colored in my eyebrows, put on liner and mascara, applied concealer around my eyes, and blushed up my cheeks and lips. When I didn't wear make-up, people would comment that I looked tired.

My weight and size have been contentious, and probably the most deleterious on my health. By the time I was in middle school, I was 5'10" and towering over most of the boys in my class. I became very conscious of this, and tried to make myself smaller. I adopted a posture where I would stand on one bent leg (my shorter one) and jet the other leg out. I would also hunch my shoulders and parallel my neck to the ground to lose an inch or two.

I was never super heavy, but always conscious of my weight. I had come to terms with being a tall woman, but I desperately didn't want to cross the line into being a "big woman." I worked out almost every day in high school and college, getting to the point where I would feel "fat" on days that I didn't. I loved food, but became anxious about it, sometimes having panic attacks at the idea of how a meal would make me look and feel. For close to two decades, I was in a perpetual state of sucking in my stomach.

And then there's periods, birth control, and pregnancy. I was put on birth control for acne, which messed with my hormones and made me a walking box of tears. I swore I would never use hormonal birth control again. When I did start having sex, I used condoms at first – but I soon learned that when you ask most men to use a condom, they look at you like you just shot their dog. I tried a non-hormonal IUD, but this gave me 12-day-long periods that were monstrously heavy and painful. I had to put in three tampons (yes, at the same time) just to get through service.

If a woman has children, the issues of poor maternal leave, no job security, health care, and child care come into play. Unless women have an incredible support system, a lot of money, or both, the prospect of maintaining an ambitious career seems out of the question. If women "just" want to stay home and be mothers, they are often vilified. And if women don't want to have children, they're labeled as selfish and unwomanly.

As women, growing up we are conditioned to be beautiful, thin, gracious, and, to a certain extent, small. Neither our voices nor our bodies should be loud. Due to this pressure, we develop afflictions that are almost exclusively ours: eating disorders, depression, anxiety, malnutrition. In turn, our bodies suffer and we get digestive syndromes, panic attacks, mood disorders, and menstrual problems. What would our lives – and our bodies– be like if we didn't suck in our stomachs, slouch to be smaller, limit our calories, or over-exercise?

There's so much more. This is a world of female genital mutilation, date rape, forced marriage, marital rape, Brock Turner (and the judge who reduced his sentence), sex-slave trade, violence against sex workers, "binders full of women," slut-shaming, Harvey Weinstein, female infanticide (as well as sex-selective abortion), the gender pay gap, restricted land ownership, dowries, victim-blaming, lack of access to health care, lack of access to education, lack of access to legal rights (including divorce), legislation over a woman's body, honor killings, domestic violence, body-shaming, statute of limitations on rape, and "grab them by the pussy" – among countless other crimes and injustices.

And despite all this, women have overcome, persevered, and achieved.

We all make choices every day that create the world we live in, for ourselves and for others. Let's strive to make kind and enlightened ones.

259

Somm Survey:
What's the hardest thing about being a woman in the wine industry? What's the hardest thing about being a woman in the world?

260

"It takes longer to earn your stripes. Your opinions or commentary can be ignored. Many times, especially among professionals, men would talk over me, correct me, interrupt, or put themselves center stage. I have struggled with confidence because of this. My advice: be confident, keep pushing, be determined."
Kara Maisano

"I remember getting tons of comments – some admiringly, some in disbelief – from co-workers (sometimes guests, even) about how on earth I managed/endured wearing three-inch heels on the floor during service. For me, it was easy. I had to. There will always be a double standard and I'm okay with that. I know many will bitch about equality, but for me, I'm okay working a little bit more than my usual 100% to be seen as an equal."
Paula de Pano

"When I was training as a wine runner and cellar master in my first restaurant job, my friend (another young, eager female assistant sommelier) and I were pulled aside right before opening night and asked if we would be cocktail servers…We were told we were selected because of our high scores on our cocktail quizzes. When we were later asked to ditch our usual black dress-pants and white button-up shirt for a small, tight black dress, I knew it wasn't because of our cocktail knowledge…"
Hannah Williams

"Any problems I've encountered in our industry are the same systemic problems that exist in *any* industry: lack of equal pay, lack of equal job opportunities, lack of parental leave, sexual harassment."
Carla Rza Betts

"People assume that because I am a woman, I don't know what I'm talking about. Or they are more critical of my appearance…When I use the same directness and straightforwardness of a man, I am criticized for it or labeled a bitch."
Christy Fuhrman

"On the winemaking and vineyard work side, I've had some pretty awful experiences, ranging from low-level 'you're not strong enough for that' or 'I just think [some dude] would be better suited to that,' to full-blown sexual harassment. You can't really complain about it, though; otherwise you're that woman who can't deal. Wine is a small community, and making a fuss about something being egregiously unfair won't get you anywhere. Just proving that you're great at what you do will shut up most people."
Claire Hill

"The lack of pockets in women's professional clothing. Having people treat what you say differently because you are a woman, i.e. I have said the EXACT same thing to staff as male colleagues but am perceived as bitchy instead of direct."
Jill Zimorski

"The constant mansplaining. It's beyond irritating and honestly insulting. In general, I think the hardest thing about being a woman in the world is worrying for your physical safety all the time."
Kelli White

"Instances of male guests not respecting me as their wine captain ('Oh, you're the sommelier? Do you have a *head* sommelier?') or flat-out grabbing me or my co-workers. Am I allowed to talk about periods? It's obviously difficult to work a 10 to 14-hour day on the floor with massive cramps and menstrual pain, not to mention dealing with heightened emotions in an already stressful workplace."
Rebecca Flynn

"Garnering the same level of respect, trust, and opportunities as men. The abuse of women and minorities in the hospitality industry is rampant. There are so few women who hold positions of power – restaurateurs, beverage directors, partners – which means that women's rights are decided by men. The only way to change the industry and to change the world is for women to invest in women. Mentor a young woman you see potential in, build up her career, have her learn from your mistakes."
Victoria James

"One of the most difficult things about being a woman in the wine industry is believing that this industry is a meritocracy. Getting people to look past your gender and to simply trust your expertise and experience can be difficult, especially in the beginning. I can easily extrapolate this to living in the world as a woman: we get less trust, which we then internalize, and our self-confidence is destroyed. No one likes a loud-mouthed broad telling them she deserves to be treated better. I'm always that broad."
Cara Higgins

The Perfect (and Fictional) Women's Sommelier Suit

Flexibility in the arms and shoulders to reach up high in the cellar and across a large table.

Shirt buttons sit flat and don't gape.

Sleeves hit right at the wrist bone, not too long or too short.

Shirt stays tucked in throughout service.

POCKETS. Lots of them. Both hips, rear pants, both jacket flaps, and at least one more inside the jacket.

Pants are made with some acknowledgment that hips exist.

Hem hits perfectly at the ankle or extends over the shoes to just hover off the floor.

Shoes are supportive, comfortable, and polishable without looking like something your grandma wears.

VIGNETTE

GRAMONA *LA CUVEE BRUT GRAN RESERVA* Cava, Spain
VAL DELLE CORTI Chianti Classico, Tuscany, Italy
G.B. BURLOTTO *ELATIS VINO ROSATO* Italy

Value Wine

The year leading up to theory was a marathon.

There are no quick sprints in this race, no shortcuts. While I don't find it arduous to study hard, the pain in my pursuit came from the anticipation of feeling worse as the exam neared, the anxiety about whether I would be able to perform on the day, and the stress of still not knowing how my health and MS journeys would play themselves out.

I felt completely out of control. And I countered this feeling by hitting the books. It was the one variable I *could* control. So I studied and studied and studied. And when I thought I had studied enough, I studied some more.

At a certain point, though, maybe it was four or five months out, I knew I couldn't keep ignoring my mental state. So I decided to meditate. I didn't know exactly what it meant or why it would help, but everyone kept talking about it. Every wellness article I read, every Master I heard talk, every doctor I saw: *you should meditate.* And I *really* didn't want to (which should have been a clue that I needed to). Nothing seemed more pointless than taking time out of my busy day, with the MS theory exam right around the corner, and sitting still for 20 minutes. Not thinking about anything. Breathing. *What the fuck.*

That first day, I sat in a chair with my two feet on the ground. This alone felt difficult. I was generally comfortable crossing my legs or having one up on the seat. Having my two feet on the ground gave me a hollow, tingly feeling in my legs.

I sat up straight, which was also difficult. My back hurt. I didn't know how to let my shoulders hang. My stomach felt tense and awkward.

I started to breathe. I practiced in and out through my nose, but I found myself struggling for air. I didn't know which parts of my body to contract and which to expand. My chest felt tight and my diaphragm was hard.

I barely made it two minutes before collapsing, my chest falling to my thighs. I hugged my legs and gasped for air. I was expecting meditation to be boring. I was expecting to be easily distracted. But to be nearly in the fits of a panic attack? I was not expecting that. I had been breathing and sitting for over 30 years – *how could I be so bad at it?*

I was about to write off the whole thing – *maybe this just isn't for me* – when it occurred to me that there might be something to the fact that I found it so difficult. Maybe I had been breathing wrong. Maybe I'd been sitting wrong. Maybe I'd been holding my shoulders wrong. Maybe I'd been holding my stomach wrong.

I started by looking into my breath. *How is one supposed to breathe?* I had never thought to look into it. And, as it turned out, I had been doing it pretty poorly. Growing up tall, I never stood solidly with two feet on the ground, allowing my spine to be long and my shoulders to drop down and back. I had held in my stomach for so long that I didn't even realize I was doing it. And when I sat there meditating that day, trying to release my stomach felt so unnatural and uncomfortable. Combined, these habits created poor breathing hygiene. I would only breathe shallowly and through my chest. I also realized that I breathed through my mouth a lot, especially when I was sleeping or exercising. This is a big no-no, with a symptom list including congestion, excess mucus production, poor sleep quality, excess lactic acid in the blood, more pollutants in the body, and even structural changes to the shape of the face.

I started to think there could be something to this. That perhaps my breathing had been having a detrimental effect on my health. So I breathed. I meditated. And, most importantly, I figured out what meditation meant to me. It wasn't about being still in a chair and not thinking about anything. It was about paying attention to what was going on in my body: being curious, not judgemental or reactionary.

VIGNETTE

And here was the big revelation: It had never occurred to me that I might be able to help myself heal. I had been to dozens of doctors and had asked each of them to help me. I figured the solution would come someday in an obscure test or an experimental drug. Or perhaps it would be a dietary or lifestyle change a therapist turned me onto. But I never thought that maybe, if I paid attention to my body, I could gain some insight into what ailed me.

I supplemented meditation with yoga, pranayama (the breathing part of yoga), salt baths, massage, work with a chiropractor, rolfing (look it up), organ massage (what it sounds like), and – perhaps the greatest discovery of all – mouth glue. I literally glued my lips together while I slept to ensure I would breathe through my nose.

And life changed, slowly but surely. I woke up more rested, less often in panic. I needed less sleep at night. I had more energy during the day. I was able to enjoy wine more, not feeling anxious or fatigued every time I had a glass. I felt more myself.

—

Jon and I had always been high-rollers when it came to buying wine. Wine was what we loved, and we were willing to spend money on it. We had few qualms about ordering expensive Burgundy, Barolo, and Riesling in restaurants, and we even built a fairly substantial cellar in New York. In the midst of this year, though – the year of paying for three sections of the MS exam, two plane tickets across the world to take those sections, visits to a naturopath, chiropractor, rolfer (probably not a word), and several masseuses, plus, to cap it all off, a WEDDING – we could not afford to be fast and loose with money. Instead, we found the pleasures of *value wine*. We drank Cava instead of Champagne, Chianti Classico instead of Brunello di Montalcino, and Nebbiolo rosé instead of Barolo. We went hard on local wine, which is especially cost-effective in Australia, and drank Barossa Grenache, Canberra Syrah, and Canterbury Pinot Blanc. I found so much joy, honesty, and freshness in these wines that I didn't even miss the more expensive bottles.

The months leading up to theory 2018 were filled with flash cards and test anxiety, but they were also some of the most exciting of my life. I had hope for the first time in years that I could feel better. I also had hope that I wouldn't spend the rest of my life being held hostage by a pricey wine addiction. And for both those things, I was grateful.

VALUE WINE

Somm Survey: What styles do you think offer the greatest value in the world of wine?

"Sherry, Verdicchio, Hunter Sémillon, Chinon, Cru Beaujolais, Naoussa, Vinho Verde, McLaren Vale Grenache, Rosso Conero, Douro reds."
Kara Maisano

"Northwest Spain (Godello, Mencía, etc.), Muscadet, Jerez, Savoie."
Christopher Bates

"Fortified or botrytis-affected sweet wines are the greatest value, if you talk about value in terms of the cost of production compared to the sale price."
Benjamin Hasko

"My favorite place for value is Beaujolais. Great price, steadily increasing quality, and a range of producers to choose from."
Cedric Nicaise

"Special Club Champagne."
Caleb Ganzer

"Mosel Riesling, Franciacorta, Northern Rhône, and Rioja."
Eric S. Crane

"Cremants from France, South American wines."
Vincent Morrow

"Chianti Classico is a heck of a bang-for-the-buck appellation. I still feel Chablis offers tremendous value, and you can get some awesome old Rioja for a great deal, and Nebbiolo in its many forms is always a good value (for the most part)."
Jeffrey Porter

"The 'little' wines from great producers: Petit Chablis from Dauvissat or Raveneau, Aligoté from Lafarge, and wines from really great co-ops like Domaine Wachau and Produttori del Barbaresco."
Jill Zimorski

"Chianti Classico, Greece, and Portugal."
Jonathan Ross

"Loire Valley sparkling and whites, Canary Islands, Sicily, anything made by Uli Stein and grower Champagnes (Savart, Bérêche, Pierre Peters, Chartogne Taillet, Agrapart in particular)."
Jordan Salcito

"Lesser-known varietals from less toney places. We regularly drink things like Gringet and Mondeuse from France, Grenache (red and white) from France and Australia, all the Portuguese white grapes, and things like Trousseau and Zinfandel from California."
Richard Rza Betts

M. CHAPOUTIER *CHANTE-ALOUETTE* Hermitage, Rhône Valley, France
GEORGES VERNAY *COTEAU DE VERNON* Condrieu, Rhône Valley, France
CHÂTEAU RAYAS Châteauneuf-du-Pape, Rhône Valley, France

Rhône Valley Whites

The week that followed theory was a dream.

Everything I'd been working towards and hoping for came to fruition. I had passed and, on top of that, I had felt good doing it. I had slept the night before. I had walked into the room clear and confident. And I delivered a passing performance. The week after, I slept through the night. I felt engaged and present each day. The physical symptoms of my severe anxiety seemed to be a distant memory. I was grateful and happy and, for the first time in years, I had hope that I was turning a corner. *This is the end of this chapter*, I thought. *I will pass practical and tasting in September. And I will feel good doing it. My life begins now.*

To go from this mindset to feeling like I was back at square one was a devastation that defies words. It was abrupt and severe. About a week after I was back in Melbourne, I woke up one day and everything was different. I thought it was a hiccup at first, a weird reaction to a new therapy; but as one day became two and two days became six and six days became 14, I could no longer think this was temporary. Instead, I came to believe that feeling good had been an anomaly, a brief blip of calm in the landscape of sickness.

I was worse than I had been in years. The approaching exams did not help matters, certainly. I had nightmares almost every night. They weren't about bombing tasting or dropping a tray of glasses in practical. They were always about the morning of, when I was waiting to go into the exam. The unbearable pounding of my heart, the sensations of weakness and faintness, the uncontrollable shaking.

I amped up my relaxation regimen: meditation, yoga, massage, flotation tanks, aromatherapy, magnesium oil, salt baths, chiropractic work. I was still on edge. Suppressing tears seemed a way of life. I would let them go in the mornings and at night when no one was around but Jon. He was just as heartbroken as I was. He saw me happy and confident and hopeful, and he saw me mere weeks later, shattered.

VIGNETTE

I was still tasting well, though. And I set up mock service scenarios for myself: practicing pouring out of a jeroboam, pouring a magnum of Champagne into tiny flutes, double decanting a magnum, etc. I was competent at these when I was by myself, or even with Jon, but the idea of doing any of it in front of two MSes at the exam seemed utterly impossible. I tried to visualize myself calm and confident, but images of me shaking so hard I couldn't perform basic tasks crowded my consciousness. Jon assured me that I looked great, and I was doing great, and I *would do great*. The images of self-doubt were fears, not realties.

To put this theory to the test, I wanted to perform in some high-pressure scenarios before I sat the exam. Jon set up a tasting with two MSes in Sydney. One who I knew a little bit, and the other who I didn't know at all. I was fine that day. Not calm, but not freaking out too much either. When I got in the room with them, I could feel my limbs become shaky. Weak. A little dizzy. *You'll be fine when you get started*, I told myself.

Twenty-five minutes can go amazingly fast when you have six wines to nail. I bombed the whites. I usually bank on my ability to assess structure, and it completely eluded me on this day. It was as if acid and alcohol became one entity, and I couldn't tell if a wine felt flat because it had low alcohol or low acid. When they told me I had called Condrieu (a high-alcohol, low-acid wine) as being Hunter Sémillon (a low-alcohol, high-acid wine), I lost it. I sat there with tears streaming down my face for the rest of the feedback. They were compassionate and gave me advice about being confident and getting into the right mental state.

I knew that with a calm and clear mind, I could have nailed those wines. And I was mad at myself for breaking down so completely. I had always prided myself on seeming put together, even when I was struggling inside. And now my vulnerabilities had been laid out bare for these two men to see. I was sure they thought I was a hysterical disaster. At best, I thought, they pitied me.

I went to the bathroom to compose myself. Once I had stopped crying, which took several minutes, I cleaned the mascara off my face. I patted my puffy red cheeks with a cold paper towel. I was sweaty and damp, and I smelled. I don't sweat or smell much in general, but when I get severely nervous, my body lets off a terrible,

269

wet odor. Maybe some primitive signal to the rest of the monkeys that we are in danger. It didn't seem to serve me in this instance.

The worst had happened. Sure, I could have missed more wines – I got all three of the reds right, a turnaround that I never patted myself on the back for – but it was most certainly not a pass. And on top of that, I couldn't trust my emotional and mental strength. My thoughts pounded in my head, constantly: *You can't do this. You cannot do this. There is no way you can do this.*

I thought about backing out of the exam. Surely I could get some doctor somewhere to write me a note. When crossing the street, I fantasized about a car hitting me. A broken leg would be enough to defer my exams for a year.

Passing seemed like it was no longer on the table. Before that day, I had convinced myself it was a possibility. That despite everything, I could still perform on the day and showcase my skills. I had believed that I was a Master Sommelier. All I needed to do was show up and perform. These thoughts were gone, replaced with thoughts of mere survival. *How can I get through the next couple weeks without having a breakdown? How can I get through the exam with a small piece of dignity? How can I feel better again?* This last question was the really troubling one. The real victory, the real turning point in my life, would not be passing this exam. It would be figuring out my health and getting to a point where I embraced life rather than feared it.

I talked to my sister, and I was reminded of the pact we'd made about a year before when she was having trouble conceiving a child. I told her I was sure it would all work out for her. That one way or the other, she would have a beautiful family that she loved. And she told me she was sure I would pass this exam, and that I would figure out my health problems. We would go on a vacation in five years, and whoever "lost" would pay for the trip.

Beth was due to have her first baby a week before the exam. I was overjoyed for her and her husband, Randy. I never mentioned the pact. It felt petty to say, "See, I was right! I said that everything would work out for you, and it has. And look at me! Still suffering. Still struggling. Still a mess." It was both self-pitying and mean. And although I didn't say it, I couldn't help thinking it.

Beth shared two pieces of advice her birthing coach had given her. She didn't intend to apply them to me, but I couldn't help making the connection.

The only way out is through.

Anxiety makes you want to recede, run away, step back. I had this impulse every second of every day leading up to the exam. But I knew I wouldn't back out. It was like I was back on that cliff in Capri contemplating the jellyfish in the sea below: the greater pain would be walking away. I knew I would take the exam, and when I accepted this fact – that the only way out was through – I became less paralyzed and more propelled.

The pain cannot be greater than you because it is you.

The anxiety was a part of me and, as such, it could not consume me. I had believed the anxiety was greater than me, that it would disable me, that it was eating me alive. But hearing this statement made me rethink things. Anxiety exists inside us as a tool to help us. The only way the anxiety could consume me was if I fed it by fighting against it. If I accepted it as a normal and helpful part of the process, I knew I could coexist with it.

271

—

Both Beth and I had some serious labor to go through in the coming weeks. Hers would result in a human life coming into the world. Mine could result in me reclaiming my own.

White Grapes of the Rhône

The divide between the red wines of the Northern and Southern Rhône is extreme. The Syrah-based wines in the north are dark and structured, with modest alcohol and flamboyance, and often quite a bit of rusticity. In the south, Grenache reigns supreme and the wines take on a greater lushness of structure and opulence of fruit. While the preferred white grapes change from the north to the south, they have a more similar feel than the reds. The white wines of the Rhône are largely defined by five things: elevated alcohol, diminished acidity, **phenolic bitterness**, viscous mouthfeel, and compatibility with oak aging. But there are subtle differences that result in some profound individual identities for these wines.

	VIOGNIER	MARSANNE
Appellations	Condrieu, in the Northern Rhône, is the premier appellation for Viognier. Viognier is also used as a blending partner in the red wines of Côte Rôtie.	Hermitage is the premier appellation for Marsanne, though Saint-Joseph, Crozes-Hermitage, and Saint-Peray all feature the grape. Though the laws of these appellations allow for Roussanne to be dominant, in application this is seldom the case.
Flavor Profile	Viognier is fat and oily, highly **terpendic** (floral, almost hoppy), low in acid, high in alcohol, and slightly bitter. Classic versions wear new oak.	Marsanne is richly textured and waxy, with high alcohol, lower acid, and phenolic bitterness. Its flavors are subtle, notes of marzipan and apricot, that marry well with moderate amounts of new oak.
Premier Wine	Georges Vernay *Coteau de Vernon*, Condrieu	Domaine Jean-Louis Chave, Hermitage (or M. Chapoutier *Chante-Alouette* for a more available option)
Value Example	Yves Cuilleron *Les Vignes d'à Côté Viognier*, Collines Rhodaniennes	Domaine Alain Graillot, Crozes-Hermitage

VIGNETTE

ROUSSANNE	GRENACHE BLANC	CLAIRETTE
Roussanne is the white grape that traverses the Northern and Southern Rhône valleys, making appearances in Saint-Joseph, Crozes-Hermitage, and Saint-Peray in the north, and Châteauneuf-du-Pape and Côtes du Rhône in the south.	Not allowed in any appellations of the Northern Rhône, Grenache Blanc holds court in all appellations that allow for white wine in the Southern Rhône. It is most often blended and rarely varietal; it is one of the 13 grapes allowed for production in Châteauneuf-du-Pape.	Clairette is also allowed in Châteauneuf-du-Pape and tends to feature prominently in the white wines of the region. Like Grenache Blanc, it is allowed in white wine appellations across the Southern Rhône, though is not seen as widely.
Roussanne and Marsanne share similar flavor profiles and structures. Roussanne is a bit more tempramental – prone to powdery mildew, wind damage, and uneven ripening – and thus less utilized. It shines in milder-climate pockets, with notes of pear, white peach, and honeysuckle.	Grenache Blanc is similar in structure to the grapes of the Northern Rhône, though with a touch more acidity and less bitterness. Its prominent flavors include white plum, lillies, and aniseed, and can assimilate well with moderate amounts of new oak.	Clairette is leaner-textured than most of the white grapes of the Rhône Valley. It still only has moderate acidity, but much more freshness. Historically it was prone to oxidation, but with modern-day winemaking its herbaceous and mineral character is able to shine through.
Château de Beaucastel *Vieilles Vignes*, Châteauneuf-du-Pape	Château Rayas, Châteauneuf-du-Pape	Domaine du Vieux Télégraphe *La Crau*, Châteauneuf-du-Pape
Éric Texier *Brézème Roussanne*, Côtes du Rhône	Rotem & Mounir Saouma *Inopia*, Côtes du Rhône Villages	Château de Montfaucon *Vin de Mme. la Comtesse*, Côtes du Rhône

RHÔNE VALLEY WHITES

ANDRÉ ET MICHEL QUENARD *LES ABYMES* Savoie, France
CAVE CALOZ *LES BERNUNES HEIDA PAÏEN* Valais, Switzerland
LES CRÊTES *PETITE ARVINE* Valle d'Aostá, Italy

274

Alpine Wine

We were in Piedmont for a friend's wedding when we found out that Penny had been born.

It was a great moment for the family, and I *knew* – I could feel across the world – the relief and joy beaming from Beth and Randy.

We drank Timorasso that day at lunch, an obscure Piedmontese white grape. The restaurant boasted an expansive view of Barbaresco, as well as sights of the Alps in the background. It was grandiose and formal, but no Barbaresco or Barolo that we had for the rest of the meal compared with the Timorasso. It was slightly nutty and herbal, with a waxiness to the texture, and still-bright acidity. It felt humble, but balanced and expressive, and perfectly comfortable in its own skin.

I talked to Beth later that day. The birth hadn't been easy, but it had been different to what she'd predicted – which I found oddly comforting. No matter what you prepare for, things end up different in some way – sometimes better, sometimes worse, but always different.

Later, at the wedding, I had a wine that reminded me of the Timorasso, at least in spirit. It was a Jacquere from Savoie, a true Alpine wine. Alpine wines – from Savoie, Switzerland, Valle d'Aosta – are rarely exported. It's as if the people from those regions know the secret and don't want it to get out. Just like the Timorasso, these wines are nutty, sometimes slightly oxidative, herbal, occasionally oaked, textural, and still maintain brightness. They are made with a modern consciousness that can conjure a simpler time. It was something my mind couldn't quite comprehend, and therefore the wines were never exactly what I anticipated. They were always better than the memory of them promised.

As I dozed off on the ride back to our hotel, I started thinking about Beth again. I originally couldn't understand why she was so nervous about going through labor. All she had to do was endure some pain. But the more I thought about it, the more I realized my situation wasn't so different. I had to live in anticipation of going through potentially a great deal of emotional and physical pain, just like Beth. I had to perform through that pain, and so did she. Perhaps I had a greater risk of public embarrassment, though I was probably overestimating this. And she certainly had the much greater risk of physical harm, for her and her baby. As dire as my situation felt, her consequences were actually much greater. There was no hierarchy of pain. And if there was, mine might very well be pretty low on the totem pole.

I woke up the next morning with these realizations crystalized. When I added up all my fears, not passing the exam was pretty low on the list. The emotional pain caused by the exam was much more troubling to me, and when I realized that all I had to do was endure this, passing the test came much more to the forefront.

On the way to the airport, I picked up my books again. I studied. I visualized myself answering questions in practical. I visualized myself putting glasses on a tray, decanting, pouring with ease and grace. I visualized myself dissecting wines during tasting in a calm and clear manner. I visualized being told that I passed. These exercises were not always easy – anxiety bubbled under the surface of each scenario. But I acknowledged and accepted the anxiety rather than trying to suppress it. I took these moments to surrender rather than clench.

These visualizations were also difficult because I felt like I was somehow not preparing myself for the worst. And I felt that there was an arrogance in visualizing myself passing. These hurdles were perhaps even more difficult to overcome than the anxiety itself. *But let's just give it a try,* I thought. If it's a disaster, I can always return to my old ways: visualizing the worst-case scenario so I can be prepared for it if it happens. I reminded myself that nothing will ever happen in exactly the way you prepare for – I'll never be able to prepare for every worst-case scenario – so I might as well visualize the best one.

I continued like this — positive, still, focused, quiet, resolved — for the days leading up to the exam. If I had thought the patience required for the multiple-day Advanced exam was hard, this was that experience on steroids. I didn't sleep well at all. My stomach was a wreck. But I didn't veer the course. I knew I was ready.

276

An Aerial View of the Alps

The alps provide a picturesque and character-defining backdrop for wine regions in France, Switzerland, Italy and beyond. The extremes in climate create extremes in flavor profile: these wines are textural, nutty, and flesh-filled, but with bright acidity and staunch structure.

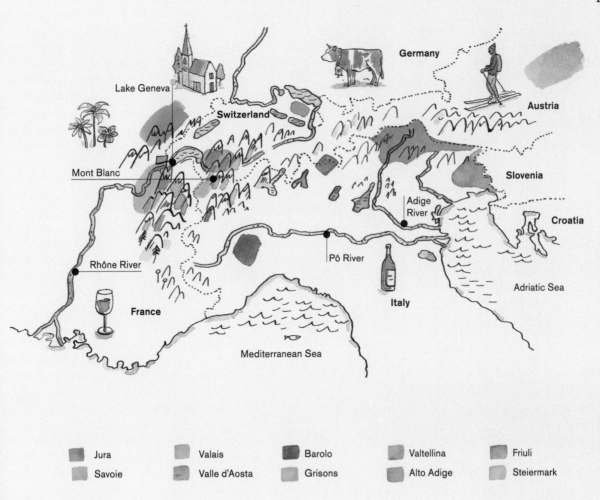

Jura	Valais	Barolo	Valtellina	Friuli
Savoie	Valle d'Aosta	Grisons	Alto Adige	Steiermark

ALPINE WINE

Champagne, Again

Before I went into each exam section that week, I did power poses. I danced around the hotel room. I gave myself fully to the moment: the anxiety, the pain, the achievement, and the possibility. And I repeated the following over and over to myself:

The only way out is through.

The anxiety cannot be greater than
you because it is you.

Don't fight the anxiety. Let it be there.

Breathe.

Let them see you shake.

Commit to being present
in every second. Don't pull
away, pull back, or try to be
somewhere else.

Surrender.

Unclench.

Smile.

Have fun.

Lead with your heart.

You've done the work. Now be yourself.
Trust yourself. And show them who you are.

CHAMPAGNE, AGAIN

Somm Survey:
On endings, what is the best way to finish a bottle of wine?

"To pause before you finish the wine and embrace the moment. I often think of the line from the film *American Beauty*: 'It's hard to stay mad when there's so much beauty in the world.'"
Pete Bothwell

"Your favorite cheeses, dark chocolate, and sex. In no particular order."
Kara Maisano

"In the bath, on a night off. Billie Holiday playing in the background, with your cat at your side."
Hannah Williams

"With a final toast to the ones we are fortunate enough to share the bottle with."
Vincent Morrow

"I always like tasting the last of a wine the next morning with a clear head."
Christopher Bates

"On a beach in Hawaii watching the sunset with loved ones."
Chris Ramelb

"With excellent company, creating a memorable (or forgettable) experience."
Rachel Speckan

"Butt chug or spoofing…just kidding. The best way is always with the people you love."
Jeffrey Porter

"With opening another. The commitment to sharing another bottle with friends and family shows the beauty and uniqueness of wine. Its ability to bring people together. To lubricate our social interaction in a way no other beverage (alcoholic or otherwise) can achieve."
Benjamin Hasko

"The best way to finish a great bottle of wine is with your closest friends already sipping on the next one."
Desmond Echavarrie

"There's no better way to finish a bottle of wine than to consider the glasses already primed for the next bottle. Emptying a bottle is a state of mind. Opening a bottle is a way of life."
Eric S. Crane

"Pour the last bit for your guests, then launch the bottle over your shoulder. While this is generally a faux shoulder toss due to safety issues, when you *can* actually complete the movement, you feel like a champion."
Carla Rza Betts

CHAMPAGE, AGAIN

Postscript

On 10 October 2018, two days after I finished the first draft of this manuscript, I woke up alone in my bed a little after 8am.

Jon had awoken at 6am that day to take care of a few things. I reached for my phone and peered at its screen through a haze of crusty eyelashes and sleepiness. There were a few texts from Jon. And one from Cameron Douglas, a New Zealand-based Master Sommelier and a member of the CMS-Americas Board of Directors. But the first thing that caught my eye was a WhatsApp message from a friend in New York: "What's going on with the CMS, are you still a Master Sommelier?"

I read the message from Jon, begging me to not check my email, to talk to Cam first. Cam reiterated the same sentiment. And of course, I checked my email. An email had gone out to the entire CMS-Americas database, anyone who had taken a Court exam or signed up for information. An email had also been sent to the entire membership of the CMS-Americas (all the Masters), of which I was still apparently a part. And a press release had been posted on GuildSomm, as well as the Facebook page of CMS-Americas: "The Board of Directors of the Court of Master Sommeliers, Americas unanimously voted to invalidate the results of the tasting portion of the 2018 Master Sommelier Diploma Examination for all candidates due to clear evidence that a Master breached the confidentiality with respect to the wines presented for tasting."

I immediately broke into tears. I had so many questions. *How could this happen? Am I still a Master Sommelier? And – most pressing – why in the world would all results be invalidated?* In the days to follow, only one was answered. I was given written notice, per the by-laws of the organization, that my membership in the Court would be suspended in 30 days.

—

On 5 September 2018, I was told that I'd passed the practical and tasting portions of the Master Sommelier exam. Many times, I had mentally rehearsed what it would feel like to get this news. In my visualizations (and dreams) Christopher Bates had given me the results. And Geoff Kruth. And Shayn Bjornholm. The Pope and the Dalai Lama might have been there too. And each time I imagined it, there were tears. I did not *imagine* tears, but I actually cried as I imagined passing, the thought was so overwhelming.

When Michael McNeill sat me down, I didn't know what to expect. I figured that if I had passed there would be a bigger to-do. When Jon was given his results, several Masters tagged along. They knew a spark of positivity was about to ignite in an otherwise gloomy day of results, and everyone wanted to be a part of it. When Michael sat down with me solo, I assumed it was a bad thing. But soon, without any lead in, he simply said: "Congratulations. You're a Master Sommelier."

I think the first thing I said was "fuck me" and threw my face into my hands. I looked up at him and waited for the flood of tears. For my emotions to come to the fore. I could feel the stirring in my chest, and just a dab of wetness in my eyes, but I was not overwhelmed. I was looking to push the emotions up, not dial them down, and I knew this was a problem. *Where are they? Why am I not losing it? Why doesn't this feel the way I thought it would?*

I thought it was a numbers issue. By the time I was called in for my results, at least a dozen people had been announced as Master Sommeliers. And almost a dozen more came after me. At the risk of sounding flip, it felt like that *Oprah* episode: "You get a pin, and you get a pin. EVERYONE GETS A PIN!" We were celebrated as the most successful year in Master Sommelier history, but it lacked climactic force for me. Perhaps I had just been conditioned to believe that this accreditation was only worth having if it was exceedingly rare.

———

In the weeks that followed the announcement that we would be stripped of our MS titles, the story of the "breach" emerged in pieces. We found out that a senior Master Sommelier and member of the CMS-Americas Board of Directors wrote an email on the morning of the exam. The recipients were blind copied, but by several accounts, it was sent to four people, two of whom later became Master Sommeliers. The email revealed two of the wines that were in the flight. The Board neither presented nor claimed that there was evidence suggesting that the confidential information had gone beyond the recipients of the email.

In this time, I became close with a group of new and more senior Master Sommeliers who questioned the Board's decision. This group reached out to several examination bodies and prestigious universities about how they would (and do) handle such situations. Not a single one said that they would universally invalidate results. The reasoned action, they all said, was to perform a thorough investigation, taking action against those involved, and exonerating fully those who there was no evidence against.

But the Board members clung to their statement that "this was the only way." They regurgitated it so frequently and matter-of-factly that it made me believe even more strongly that it was *not* the only way. I asked for support, guidance, proof, logic, reasoning, and discourse, and all I got was this threadbare phrase. *This was the only way.*

As if this phrase would somehow help me explain to my publishers why they couldn't put "MS" after my name on my book cover. As if it would give me the money and time to prepare for the exam again. As if it would erase headlines like "Australia's first female Master Sommelier in cheating scandal."

Early after the news broke, I sent an email to a couple dozen Master Sommeliers who I had a personal relationship with. I urged them to think about how the Board's precedent – of being able to rescind credentials at any time – could affect them. And I asked them to consider that perhaps the better precedent for the organization to set is that it is able to investigate and ferret out wrong-doing, rather than having to invalidate entire exams.

Master Sommeliers outside the Board were just as unresponsive as those on it. Most wanted to keep their heads down and not get involved. People I considered close friends – people who'd mentored me, people I'd known for ten years, people who were invited to my wedding! – all silent. This might have been the most shocking part.

As for the 23 members of the class of 2018 who lost their pins: I can't judge how anyone responds in a situation like this. Some members of our class believe that if they are to be part of this organization, they must trust and accept the Board's decision. Others may not accept it, but fear the ramifications of speaking out. Some have been very vocal in their outrage on social media. Others haven't said a word.

As for me: I was raised to speak out against injustice. My parents – both retired lawyers – were proud of me when I passed the Master Sommelier exam. And they were just as proud when I decided to attend my suspension hearing, and make it clear to the Board of Directors of the CMS-Americas that I did not accept their decision.

At my hearing, the Board did answer some of my questions. They acknowledged they have no evidence that I cheated, but that even if I did pass fairly, my credential should be revoked. And, perhaps most importantly, they admitted that once they determined the exam had been compromised, they ceased to perform any kind of investigation.

Within a week they sent out a letter stating that no evidence presented at the hearing had swayed their course. My membership would be suspended the second week of November 2018.

After this communication, I went silent. I had to focus on planning my wedding, and on this book, which was essentially a love letter to the Court of Master Sommeliers: how much of myself I had dedicated to the organization, and my great respect for the credential and the people who had attained it. How would these new developments affect my through-line?

The first opportunity to retake the tasting exam was granted on 5 December 2018. Thirty people took the exam. Fourteen or fifteen (by different accounts) of those had passed in September, had enjoyed the title of Master Sommelier for five weeks, then were stripped of it. This time, six people passed. All who had passed in September. The fact that no one who failed in September passed three months later supports the mythology of the exam: that it is not won overnight.

285

But eight or nine people who had passed in September did not pass the retest. What, if anything, did this prove? Was the exam harder? Was the stress – tasting to prove one's innocence – a factor? Is the exam really a random test? Or were the people who sat in September privy to confidential knowledge that helped them pass (the least likely scenario in my opinion)?

—

There are some very flawed individuals in this organization. Perhaps I am one of them. There's no doubt many people have taken issue with my response to this crisis. I'm sure many viewed my initial outreach to my peers as immature and emotional. I'm sure being represented by a lawyer at the hearing was viewed as an act of aggression against the Court. And this postscript could certainly be viewed as heresy. But I do not apologize for my actions. Just as the Board will not apologize for theirs.

As I write this, I have not yet decided if I will sit the exam in 2019. Part of me feels that agreeing to a retest means I accept their course of action. But another part of me wants to reclaim the title I rightfully earned. And down the track, perhaps influence the organization from the inside.

Despite everything, I stand by my initial thesis outlined in the previous pages: that the Master Sommelier exam builds character and community like none other. And though in a very different form than I had once imagined, it has continued to build character and community in these trying times.

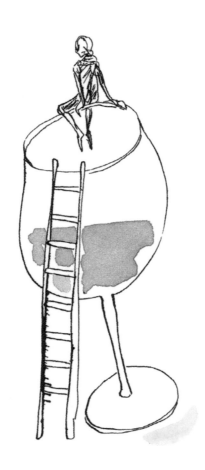

POSTSCRIPT

Glossary

AVA: American Viticultural Area, registered and protected by the Tax and Trade Bureau.

Bamboo: A cocktail of amontillado sherry, dry vermouth, and bitters. Created by Louis Eppinger at the Grand Hotel in Yokohama, Japan, in the 1890s.

Barolo Chinato: Fortified and aromatized Barolo, a common bittersweet after-dinner drink in Piedmont.

Barrique: A small barrel, approximately 225 liters in size (see page 194).

Biodynamics: A branch of viticulture based on the teachings of Rudolf Steiner. Biodynamics focuses on creating an enclosed ecosystem within a vineyard, using natural preparations to treat it, and timing actions with lunar cycles.

Brettanomyces: Brettanomyces, or "brett," is a yeast family that is common in lambic beer production. It is also commonly found in the red wines of Bordeaux, the Rhône Valley, Bandol, and Burgundy. Left unchecked, the horsey and rubbery aromatics can become unpleasant.

Cap Management: During red wine fermentations (or any fermentations where the grape skins are included), the CO_2 created pushes the skins to the top of the fermentation vessel, forming a cap. If left "unmanaged," the cap will dry out, allowing for potential oxidation and not allowing for proper extraction of the color and phenolics of the skins. The cap can be managed through a variety of different methods, including punch downs (pigéage), pump overs (remontage), or racking and returning (délestage).

Carbonic Maceration: A winemaking technique for red wines, whereby grape juice is encouraged to begin fermenting inside the skin of the grape. Can result in lower tannins, brighter color, primary fruit notes, and a pickled quality.

Chaptalization: The act of adding sugar to grape juice before it ferments. The idea is not to add sweetness to the final wine, but to add potential alcohol. This practice is not uncommon throughout classic cool-climate regions of Europe.

Climat: A term often used to designate a vineyard site that is not legally defined.

Clonal Selection: Grapes have evolved and mutated over the years, creating multiple (sometimes dozens) of clones of the same grape. Clonal selection describes the act of choosing clones to propagate in a vineyard (usually from a nursery). The opposing method is called *Selection Massale*, under which a random sample is selected from a vineyard site, allowing for clonal diversity to thrive instead of selecting for a specific clone.

Coravin: A system of wine preservation whereby a needle is inserted through a bottle's intact cork. As the wine is removed, it is replaced with argon gas, which displaces the oxygen and keeps the wine fresh.

Côte de Beaune: The southern part of the Côte d'Or (the main stretch of Burgundy), extending from just south of Nuits-Saint-Georges to the River Dheune, and including the town of Beaune.

Côte de Nuits: The northern part of the Côte d'Or (the main stretch of Burgundy), extending from Dijon to Nuits-Saint-Georges.

DOCG: Denominazione di Origine Controllata e Garantita, the highest quality designation for appellations in Italy.

Falernum: A rum-based liqueur made with lime, ginger, and clove.

Feinherb: Feinherb roughly translates to "fine tart" and is a labeling term used in Germany. It is not a legally defined term, but was created because of the poor image of *halbtrocken* wines (which are legally defined as wines between 10 and 18 grams per liter residual sugar). Feinherb wines have exhilarating balance and can be some of the greatest in Germany, with just a hint of residual sugar.

Fermentation: The process of yeast converting sugar to alcohol, with the biproducts of CO_2 and heat.

Fortification: The act of adding a distilled spirit to wine, either before, during, or after fermentation.

Grosse Lage: The top vineyard designation in Germany, translating to "great site."

Grosses Gewächs: A designation assigned by Germany's VDP, denoting dry wine from a Grosse Lage site (see above). Noted on a bottle by the abbreviation "GG."

289

Gueridon: The French word for a service cart, the traditional vehicle for tableside decanting and wine service.

Grand Cru: A term widely used across France to designate quality. It is most often associated with Burgundy, whereby Grand Crus are considered the greatest vineyard sites. The term can be used in other regions (often with less specificity and meaning).

Lees: The expended yeast cells from alcoholic fermentation, often kept in contact with the wine to create body, flavor, or consume oxygen.

Lees Handling: The way in which lees are managed post-fermentation; this often refers to how often (if at all) lees are "stirred."

Martinez: A classic cocktail made with gin, sweet vermouth, maraschino liqueur, and orange bitters. The Martinez is considered the precursor to the Martini, and was created in California in the 1860s.

McKittrick Old Fashioned: An old-fashioned made with bourbon, Pedro Ximénez sherry, and mole bitters. Created by Theo Lieberman at Milk & Honey in Manhattan, circa 2011.

MS: Shorthand for Master Sommelier.

MW: Shorthand for Master of Wine, another prestigious wine qualification, bestowed by the Institute of Masters of Wine in the UK. The Master of Wine certification is not service-based, with more writing and analytics, attracting more wine writers and winemakers than sommeliers.

Negociant: A producer who buys grapes/finished wine instead of growing/making it.

Oidium: A fungal disease caused by powdery mildew.

Organic: Viticulture that focuses on organic additives and eschews the use of chemical fertilizers, pesticides, and fungicides.

Permaculture: Sustainable agricultural systems aimed at increasing organism diversity.

Phenolic Bitterness: A term primarily used to describe the bitterness in white wines caused by phenols in their skins. White grapes with lower acidity – like Pinot Gris, Gewürztraminer, Viognier, and Marsanne – are balanced by phenolic bitterness.

Phylloxera: An aphid that feeds on the roots of grapevines.

Phylloxera devastated the vineyards of Europe in the late 19th century, and has impacted most wine regions across the world. The most effective protection against phylloxera is to graft *vitis vinifera* on native American grapevines, which are immune to the pest.

Pipeños: A word describing traditional Chilean wine, aged in *pipas* (redwood casks) and featuring traditional grapes like Pais and Moscatel.

Port-Tong: A method of removing the cork from a bottle, originally designed for Port wine. (see page 179).

Premier Cru: A term used in France to designate quality. It is most often associated with Burgundy, whereby Premier Crus are the tier below Grand Crus in vineyard quality designation.

Proof: A term used to measure the alcoholic strength of spirits. Proof is equal to double the amount of the alcohol by volume, i.e. 45% alcohol is 90 proof.

Pyrazines: A class of chemical compounds naturally occurring in certain grape varieties that express themselves as green pepper, tarragon, green bean, and grass. Most prevalent in Cabernet Sauvignon, Cabernet

Franc, Carménère, and Sauvignon Blanc (see page 212).

Racking: The act of transferring wine from one vessel to another, often used to remove a wine from its solids.

Reduction: The opposite of oxidation. It can be created through reductive winemaking techniques and can result in a *reduced* wine. Subtle reduction can be desired and is especially prevalent in the white (and to a lesser extent, red) wines of Burgundy. It can smell like flint and gun smoke, or it can veer into cabbage, hard-boiled eggs, and onion if certain unpleasant sulfides develop (see pages 40 and 212).

Residual Sugar: Shortened to "RS," residual sugar measures the sugar content (in grams per liter) in a finished wine.

Rotundone: A chemical compound naturally occurring in certain grape varieties, expressing itself as black and white pepper, and smoke. Syrah is the most common grape that contains rotundone, but others include Pineau d'Aunis, Duras, Negrette, Fer, Vespolina, Schioppettino, Mondeuse and the white grape Grüner Veltliner (see page 212).

Saignée: A red-winemaking technique also known as "bleeding the tanks" whereby a small amount of juice is "bled off" pre-fermentation after a short period of skin contact. This enhances the color and concentration of the remaining red wine. Alternatively (or concurrently), this can be a technique used to create rosé wine, where the "bled off" juice is desired.

Sekt: A term for sparkling wine in Germany and Austria.

Slavonian botti: Slavonia is a historic region in Croatia that grows a sub-species of Quercus Robur (see page 194). The wood is tight-grained and sturdy and often made into large barrels (botti) used in Italy.

Spinning Cones: A winemaking technology used to separate alcohol from other components of the wine, allowing the alcohol in a wine to be decreased without diluting flavor.

Tannin: Naturally occurring compound found in grape skins (as well as seeds, stems, leaves, and other things found in nature). This compound, which can also be synthesized by oak treatment, creates the drying mouthfeel of some red wine.

Terpenes: Chemical compounds found in certain grape varieties (mainly white) that manifest as extreme florality. Muscat, Torrontes, Gewürztraminer, Viognier, Albariño, and Riesling are all grapes that express terpenes at varying levels.

Tokaji: An appellation in northeastern Hungary famous for sweet, botrytis-influenced wines. Tokaj is the place; Tokaji is the wine/appellation.

Village: A term encompassing Burgundy wines made from a single village, but not a more specific Premier Cru or Grand Cru vineyard site.

Vintage: The year in which grapes are harvested (picked) for a particular wine. Vintage usually occurs in September and October in the northern hemisphere and March and April in the southern.

Volatile Acidity: "VA," as it is affectionately known, is common in Italian red wines, a by-product of long fermentations and macerations occurring in open-air environments. It lifts aromatics and can add a (pleasant) balsamic and acetone character. It can also crop up in wines without much sulfur protection and can become a distraction at high levels instead of a subtle enhancement.

291

Value Wines Appendix

This appendix contains value wines that represent every style of beverage featured in the book. The price point can vary with each style, but the idea is to find the lowest entry point at which a true expression of the style can still be accessed. A few chapters are not included here: repeated styles (Champagne, Barolo), categories that defy multiple listings (Green Chartreuse, Campari), and one left off for redundancy (Value Wines).

Champagne

Bérêche et Fils *Brut Réserve*,
  Champagne, France
Agrapart et Fils *7 Crus*,
  Champagne, France
Pol Roger *Réserve*,
  Champagne, France
Louis Roederer *Brut Premier*,
  Champagne, France
La Caravelle *Cuvée Niña*,
  Champagne, France

California Chardonnay

Sandhi *Chardonnay*,
  Santa Barbara County,
  California, USA
Mary Elke *Chardonnay*,
  Anderson Valley, Mendocino
  County, California, USA
Frank Family *Chardonnay*, Napa
  Valley, California, USA
Hanzell *Sebella*
  Chardonnay, Sonoma
  County, California, USA
Domaine Eden *Chardonnay*,
  Santa Cruz Mountains,
  California, USA

Rum

Plantation *3 Stars*, Jamaica/
  Barbados/Trinidad
El Dorado *12-Year*, Guyana
Barbancourt *4-Year*, Haiti
Brugal *Añejo*,
  Dominican Republic
Smith & Cross, Jamaica

Sangiovese

Gianni Brunelli, Rosso di
  Montalcino, Tuscany, Italy
Il Borghetto *Montigiano*,
  Tuscany, Italy
Il Macchione, Vino Nobile di
  Montepulciano, Tuscany, Italy
Querciabella, Chianti Classico,
  Tuscany, Italy
Le Ragnaie *Troncone*,
  Tuscany, Italy

Portuguese Wine

Herdade do Esporão *Tinto
  Reserva*, Alentejo, Portugal
Anselmo Mendes *Muros
  Antigos Loureiro*, Vinho
  Verde, Portugal
Vadio *Tinto*, Bairrada, Portugal
Beyra *Quartz Branco*, Beira
  Interior, Portugal
Filipa Pato *FP Branco*,
  Bairrada, Portugal

Scotch

Compass Box *The Spice
  Tree Blended Malt Scotch
  Whisky*, Scotland
Glenmorangie *10-Year*,
  Highlands, Scotland
Isle of Skye *8-Year*, Scotland
Monkey Shoulder, Scotland
Auchentoshan *12-Year*,
  Lowlands, Scotland

White Burgundy

Les Héritiers du Comte Lafon,
  Viré-Clessé, Mâcon,
  Burgundy, France
Domaine Méo Camuzet *Clos
  Saint-Philibert*, Bourgogne
  Hautes-Côtes du Nuits, France
Domaine Jean-Philippe Fichet,
  Bourgogne, France
Domaine A. & P. de Villaine,
  Bouzeron, Côte Chalonnaise,
  Burgundy, France
Simon Bize et Fils, Savigny-les-
  Beaune, Burgundy, France

Tequila

El Tesoro *Blanco*, Jalisco, Mexico
Cazadores *Reposado*,
  Jalisco, Mexico
Gran Centenario *Blanco*,
  Jalisco, Mexico
1800 *Añejo*, Jalisco, Mexico
Astral *Blanco*, Jalisco, Mexico

Botrytis

Domaine des Baumard *Carte
  d'Or*, Coteaux du Layon,
  Loire Valley, France
Kiràlyudvar *Cuvée Ilona*,
  Tokaji, Hungary
Château de Suduiraut *Lions
  de Suduiraut*, Sauternes,
  Bordeaux, France
De Bortoli *Noble One Botrytis
  Semillon*, Riverina, Australia
Georg Gustav Huff *Niersteiner
  Schloss Schwabsburg
  Riesling Beerenauslese*,
  Rheinhessen, Germany

American Pinot Noir

Hirsch Vineyards *San Andreas Fault Pinot Noir*, Sonoma Coast, California, USA

Brooks *Pinot Noir*, Willamette Valley, Oregon, USA

Cristom *Mt. Jefferson Cuvée Pinot Noir*, Willamette Valley, Oregon, USA

Tyler *Pinot Noir*, Santa Barbara County, California, USA

Syncline *Pinot Noir*, Columbia Gorge, Washington, USA

Austrian Wine

Prager *Hinter der Burg Grüner Veltliner Federspiel*, Wachau, Austria

Strohmeier *Schilcher Frizzante*, Steiermark, Austria

Weingut Lagler *1000 Eimerberg Neuburger*, Wachau, Austria

Prieler *Blaufränkisch*, Leithaberg, Burgenland, Austria

Fred Loimer *Gumpoldskirchner*, Thermenregion, Austria

Amari

Fratelli Branca *Fernet-Branca*, Milan, Italy

Lucano 1894 *Amaro Lucano*, Bascilicata, Italy

Montenegro *Amaro Montenegro*, Bologna, Italy

Fratelli Averna *Averna Amaro Siciliano*, Sicily, Italy

Ditta Silvio Meletti *Amaro Meletti*, Marche, Italy

Merlot

Macari *Reserve Merlot*, North Fork of Long Island, New York, USA

Venica & Venica *Merlot*, Collio, Friuli, Italy

Villa Maria *Private Bin Organic Merlot*, Hawkes Bay, New Zealand

L'Ecole No. 41 *Merlot*, Columbia Valley, Washington, USA

Château Grand Village, Bordeaux Supérieur, France

Barolo (and Other Italian Nebbiolo)

Renato Ratti *Ochetti*, Nebbiolo d'Alba, Piedmont, Italy

Cordero di Montezemolo *Monfalletto*, Barolo, Piedmont, Italy

Sandro Fay *Téi*, Rosso di Valtellina, Lombardy, Italy

Produttori del Barbaresco, Barbaresco, Piedmont, Italy

G.B. Burlotto, Langhe Nebbiolo, Piedmont, Italy

Vinos de Jerez

Valdespino *Fino Inocente*, Jerez-Xérès-Sherry, Andalusia, Spain

Bodegas Cesar Florido *Moscatel Dorado*, Jerez-Xérès-Sherry, Andalusia, Spain

Gonzalez Byass *Elegante Amontillado*, Jerez-Xérès-Sherry, Andalusia, Spain

Gutierrez Colosia *Oloroso*, Jerez-Xérès-Sherry, Andalusia, Spain

Bodegas Hidalgo *La Gitana*, Manzanilla-Sanlúcar de Barrameda, Andalusia, Spain

Kiwi Wine

Pegasus Bay *Riesling*, Waipara Valley, Canterbury, New Zealand

Te Mata *Gamay Noir*, Hawkes Bay, New Zealand

Rippon *Jeunesse Young Vine Pinot Noir*, Lake Wanaka, Central Otago, New Zealand

Craggy Range *Te Muna Road Sauvignon Blanc*, Martinborough, New Zealand

Kumeu River *Estate Chardonnay*, Kumeu, Auckland, New Zealand

Gin

Four Pillars *Rare Dry Gin*, Yarra Valley, Victoria, Australia

Beefeater *London Dry Gin*, Kennington, England

Aviation *American Gin*, Portland, Oregon, USA

Boodles *British Gin London Dry*, Warrington, England

Tanqueray *London Dry Gin*, Windygates, Scotland

Chenin Blanc

Mia et Kenji Hodgson *Faia Vin Blanc*, France

Domaine Huet *Le Haut-Lieu Sec*, Vouvray, Loire Valley, France

Domaine de Bellivière *Les Rosiers*, Jasnières, Loire Valley, France

Millton *Te Arai Vineyard Chenin Blanc*, Gisborne, New Zealand

Domaine du Closel *La Jalousie*, Savennières, Loire Valley, France

Sauvignon Blanc

Dog Point Vineyard *Sauvignon Blanc*, Marlborough, New Zealand

Clos Floridene, Graves, Bordeaux, France

Selene *Hyde Vineyard Sauvignon Blanc*, Carneros, Napa Valley, California, USA

Edi Kante *Sauvignon*, Venezia Giulia, Italy

Michel Redde *Petit Fumé*, Pouilly Fumé, Loire Valley, France

American Whiskey

Eagle Rare *10-Year Straight Bourbon Whiskey*, Frankfurt, Kentucky, USA

Woodford Reserve *Straight Bourbon Whiskey*, Versailles, Kentucky, USA

Jim Beam *Pre-Prohibition Style Rye*, Clermont, Kentucky, USA

Rittenhouse *Straight Rye Whiskey*, Bardstown, Kentucky, USA

Wild Turkey *101 Straight Bourbon Whiskey*, Lawrenceburg, Kentucky, USA

German Riesling

Dr. Bürklin-Wolf *Estate Riesling*, Pfalz, Germany

Julian Haart *Riesling*, Mosel, Germany

Horst Sauer *Escherndorfer Lump Riesling Kabinett Trocken*, Franken, Germany

A.J. Adam *Dhroner Hofberg Riesling Kabinett*, Mosel, Germany

Peter Lauer *Senior Ayler Riesling Fass 6*, Saar, Germany

Italian White Wine

Abbazia di Novacella *Sylvaner*, Alto Adige, Italy

Venica & Venica *Friulano*, Collio, Friuli, Italy

Pieropan, Soave Classico Veneto, Italy

Graci, Etna Bianco, Sicily, Italy

Sartarelli, Verdicchio dei Castelli di Jesi Classico, Marche, Italy

California Cabernet (Napa and Others)

Mount Veeder Winery *Cabernet Sauvignon*, Mount Veeder, Napa Valley, California, USA

Etude *Cabernet Sauvignon*, Napa Valley, California, USA

Ryme *Peacelands Vineyard Cabernet Sauvignon*, Fountaingrove District, Sonoma County, USA

Rodney Strong *Cabernet Sauvignon*, Sonoma County, California, USA

Brea *Margarita Vineyard Cabernet Sauvignon*, Paso Robles, California, USA

The Other 46 States

Barboursville Vineyards *Reserve Nebbiolo*, Monticello, Virginia, USA

Caduceus Cellars *Lei Li Nebbiolo Rosé*, Graham County, Arizona, USA

L. Mawby *Blanc de Blancs*, Leelanau Peninsula, Michigan, USA

McPherson Cellars *Les Copains*, Texas, USA

Snowy Peaks Winery *Elevé*, Grand Valley, Colorado, USA

Red Burgundy

Domaine Chevillon, Bourgogne Passetoutgrains, France

Domaine Fontaine-Gagnard, Bourgogne, France

Domaine Tollot-Beaut, Chorey-les-Beaune, Burgundy, France

Domaine Faiveley *La Framboisière*, Mercurey, Burgundy, France

David Duband *Louis Auguste*, Bourgogne-Hautes Côtes de Nuits, France

Cheese Wine

Etienne Dupont *Cidre Bouché Brut de Normandie*, Normandy, France

Royal Tokaji Co. *Late Harvest*, Tokaj, Hungary

Domaine la Tour Vieille *Reserva*, Banyuls, Roussillon, France

Blandy's *5-Year Bual*, Madeira, Portugal

André et Mireille Tissot, Crémant du Jura, France

Chablis

Clotilde Davenne, Chablis, Burgundy, France
Domaine Oudin *Les Serres*, Chablis, Burgundy, France
Vincent Dauvissat, Petit Chablis, Burgundy, France
Domaine Christian Moreau, Chablis, Burgundy, France
Jean Paul & Benoît Droin, Chablis, Burgundy, France

Calvados

Roger Groult *Age 8 Ans*, Calvados Pays d'Auge, Normandy, France
Lemorton *Réserve*, Calvados Domfrontais, Normandy, France
Famille Dupot *Vieille Réserve*, Calvados Pays d'Auge, Normandy, France
Christian Drouin *Domaine Coeur du Lion Sélection*, Calvados, Normandy, France
Adrien Camut *6 Ans d'Âge*, Calvados Pays d'Auge, Normandy, France

Beer

Brouwerij Rodenbach *Grand Cru*, Roeselare, Belgium
Plzeňský Prazdroj *Pilsner Urquell*, Plzeň, Czech Republic
Gasthaus & Gosebrauerei Bayerischer Bahnhof *Leipziger Gose*, Leipzig, Germany
Kirin *Ichiban*, Yokohama, Japan
Sierra Nevada Brewing Co. *Pale Ale*, Chico, California, USA

Syrah

Domaine Rostaing *Les Lézardes*, Collines Rhodaniennes, Rhône Valley, France
Mac Forbes *Syrah*, Yarra Valley, Victoria, Australia
Stolpman Vineyards *Estate Grown Syrah*, Ballard Canyon, Santa Barbara County, California, USA
Domaine Gramenon *Sierra du Sud*, Côtes du Rhône, Rhône Valley, France
Clonakilla *O'Riada Shiraz*, Canberra District, New South Wales, Australia

New York Wine

Hermann J. Wiemer *Dry Riesling*, Seneca Lake, New York, USA
Shinn Estate Vineyards *Mojo Cabernet Franc*, North Fork of Long Island, New York, USA
Chëpika *Catawba*, Finger Lakes, New York, USA
Element Winery *Pinot Noir*, Finger Lakes, New York, USA
Channing Daughters *Ramato Pinot Grigio*, Long Island, New York, USA

Bordeaux

Château le Puy *Emilien*, Francs-Côtes de Bordeaux, France
Château Cantemerle, Haut-Médoc, Bordeaux, France
Château Peybonhomme-les-Tours, Blaye-Côtes de Bordeaux, France
Château Fonroque, Saint-Émilion Grand Cru, Bordeaux, France
Château Le Bouscat, Haut-Médoc, Bordeaux, France

Port

Niepoort *Dry White*, Porto, Portugal
Symington's *Quinta do Vesuvio Vintage*, Porto, Portugal
Warre's *Late Bottled Vintage*, Porto, Portugal
Quinta do Noval *10 Year Old Tawny*, Porto, Portugal
Croft *Distinction Special Reserve*, Porto, Portugal

American-Oaked Wine

La Rioja Alta *Viña Alberdi Reserva*, Rioja, Spain
Rockford *Rod & Spur Shiraz Cabernet*, Barossa Valley, South Australia, Australia
Frog's Leap *Zinfandel*, Napa Valley, California, USA
R. López de Heredia *Viña Cubillo Crianza*, Rioja, Spain
Bodegas el Nido *Clio*, Jumilla, Murcia, Spain

Amarone (and Valpolicella)

Marion, Valpolicella Superiore, Veneto, Italy
Prà, Amarone della Valpolicella, Veneto, Italy
Masi *Costasera*, Amarone della Valpolicella Classico, Veneto, Italy

Allegrini, Amarone della
Valpolicella Classico,
Veneto, Italy
Musella, Valpolicella Superiore
Ripasso, Veneto, Italy

Irish Whiskey

Powers *Gold Label*, Midleton,
County Cork, Ireland
Redbreast *Single Pot Still
12-Year*, Midleton, County
Cork, Ireland
Mitchell & Son *Green Spot
Single Pot Still*, Midleton,
County Cork, Ireland
Knappogue Castle *Single
Malt 12-Year*, County
Antrim, Ireland
Paddy *Irish Whiskey*, Midleton,
County Cork, Ireland

Eau de Vie

Bolyhos Ágyas *Szilva Pálinka*,
Bicske, Hungary
Jean Paul Metté *Eau-de-Vie
d'Abricot*, Alsace, France
G.E. Massenez *Framboise
Sauvage*, Alsace, France
Clear Creek Distillery *Eau de Vie
of Douglas Fir*, Hood River,
Oregon, USA
Hans Reisetbauer *Williams Pear*,
Axberg, Austria

Loire Valley Reds

Domaine de Bellivière
Rouge-Gorge, Coteaux du
Loir, Loire Valley, France

Puzelat-Bonhomme *KO In Côt
We Trust*, Touraine, Loire
Valley, France
Domaine Les Hautes Noëlles
Gamay, Loire Valley, France
Catherine et Pierre Breton *Nuits
d'Ivresse*, Bourgueil, Loire
Valley, France
Jean-Max Roger *Cuvée La
Grange Dîmière*, Sancerre,
Loire Valley, France

Greek Wine

Domaine Economou *Oikonomoy
Liatiko*, Crete, Greece
Hatzidakis Winery, Santorini,
Greece
Kir-Yianni *Paranga Sparkling*,
Amynteo, Macedonia, Greece
Kokkinos *Xinomavro*, Naoussa,
Macedonia, Greece
Tselepos Winery *Moschofilero*,
Mantinia, Peloponnese, Greece

Breakfast Wine

Patrick Bottex *La Cueille*, Bugey
Cerdon, Savoie, France
Cleto Chiarli *Vigneto Cialdini*,
Lambrusco Grasparossa
di Castelvetro, Emilia-
Romagna, Italy
Giacomo Bologna *Braida*,
Brachetto d'Acqui,
Piedmont, Italy
Pizzini *Brachetto*, King Valley,
Victoria, Australia
Grant Burge *Shiraz Cabernet
Méthode Traditionelle*,
Australia

Mezcal

Del Maguey *Vida*, Oaxaca, Mexico
Sombra *Joven*, Oaxaca, Mexico
El Jolgorio *Nuestra Soledad*,
Oaxaca, Mexico
Alipús *San Juan del Rio*,
Oaxaca, Mexico
Mezcal Vago *Elote*,
Oaxaca, Mexico

Vodka

Tito's *Handmade Vodka*,
Austin, Texas, USA
Hippocampus Metropolitan
Distillery *Vodka*, Braeside,
Victoria, Australia
Blue Duck *Rare Vodka*,
Auckland, New Zealand
Żubrówka *Bison Grass Vodka*,
Brest, Belarus
Still the One Distillery *COMB
Vodka*, Port Chester,
New York, USA

Alsatian Wine

Domaine Weinbach *Réserve
Personelle Gewürztraminer*,
Alsace, France
Josmeyer *Mise du Printemps
Pinot Blanc*, Alsace, France
Domaine Paul Blanck *Pinot Gris*,
Alsace, France
Domaine Albert Mann *Muscat*,
Alsace, France
Albert Boxler *Riesling*,
Alsace, France

Aussie Wine

Bird on a Wire *Marsanne*, Yarra Valley, Victoria, Australia

Crawford River *Young Vines Riesling*, Henty, Victoria, Australia

Foster e Rocco *Sangiovese*, Heathcote, Victoria, Australia

Cirillo *The Vincent Grenache*, Barossa Valley, South Australia, Australia

Reed *Knife Edge Shiraz*, Grampians, Victoria, Australia

Natural Wine

Marcel Lapierre *Raisins Gaulois*, France

Domaine La Grange Tiphaine *Clef de Sol*, Montlouis-Sur-Loire, Loire Valley, France

Domaine Gauby *Les Calcinaires*, Côtes du Roussillon Villages, Roussillon, France

La Stoppa *Trebbiolo Rosso*, Emilia-Romagna, Italy

Patrick Sullivan *Waterskin*, Baw Baw Shire, Gippsland, Victoria, Australia

South American Wine

Bodega Catena Zapata *Catena Malbec*, Mendoza, Argentina

Baron Philippe de Rothschild *Anderra Carménère*, Maipo Valley, Chile

Terrazas de los Andes *Torrontés*, Cafayate, Salta, Argentina

Mauricio Gonzalez Carreño *Pais*, Bío Bío Valley, Chile

De Martino *Gallardía Cinsault*, Itata Valley, Chile

Grenache

Chateau Rayas *La Pialade*, Côtes du Rhône, Rhône Valley, France

Comando G *La Bruja de Rozas*, Valle del Tiétar, Sierra de Gredos, Spain

Neyers *Rossi Ranch Grenache*, Sonoma Valley, California, USA

David & Nadia *Grenache*, Swartland, Western Cape, South Africa

An Approach to Relaxation *Sucette*, Barossa Valley, South Australia, Australia

Women's Wine

Arianna Occhipinti *Il Frappato*, Sicily, Italy

Cullen *Rose Moon*, Wilyabrup, Margaret River, Western Australia, Australia

Domaine Mee Godard *Corcelette*, Morgon, Beaujolais, France

Raft Wines *Grist Vineyard Syrah*, Dry Creek Valley, Sonoma County, California, USA

Eva Fricke *Riesling*, Rheingau, Germany

Rhône Valley Whites

Domaine Jamet, Côtes du Rhône, Rhône Valley, France

Yves Cuilleron *Les Vignes d'à Côté Viognier*, Collines Rhodaniennes, Rhône Valley, France

Rotem & Mounir Saouma *Inopia*, Côtes du Rhône Villages, Rhône Valley, France

Domaine Alain Graillot, Crozes-Hermitage, Rhône Valley, France

Domaine Bernard Gripa *Les Figuiers*, Saint-Péray, Rhône Valley, France

Alpine Wine

André et Michel Quenard *Les Abymes*, Savoie, France

Cave Caloz *Les Bernunes Heïda Paien*, Valais, Switzerland

Les Crêtes *Petite Arvine*, Valle d'Aosta, Italy

Elio Ottin *Torrette Superieur*, Vallee d'Aoste, Italy

Neumeister *Steirische Klassik Welschriesling*, Vulkanland Steiermark, Austria

Sommelier Appendix

A huge thank you to the following people who contributed their time and wisdom to this book.

Christopher Bates Element Winery, Finger Lakes, New York @sommelierbates

Yannick Benjamin Wheeling Forward, New York, New York @yannickbenjamin

Pete Bothwell Continuum Estate, New York, New York @petebothwell

Jim Bube Hogsalt Hospitality, Chicago, Illinois @jbube @jimbube

Eric S. Crane Empire Distributors, Inc., Atlanta, Georgia @chauneuf

Desmond Echavarrie Scale Wine Group, Napa, California @desiechavarrie @scalewine

Jane Lopes was born in Napa, California: wine was in her blood from day one.

She graduated from the University of Chicago with a degree in Renaissance literature and a love for food, wine, and spirits. While taking a year off after college to apply to grad school, Jane found a job in wine retail. Within a few months, she abandoned her dreams of academia and decided to pursue a career in beverage.

Soon she began to pick up shifts at nationally acclaimed cocktail bar The Violet Hour, and in August of 2011, Jane moved to Nashville to be the opening beverage director at The Catbird Seat. Her unique style of beverage pairings and approach to her program were featured in the *New York Times*, *Imbibe*, and *Wine Enthusiast*. After moving to New York in 2013, Jane began working as a sommelier at Eleven Madison Park, where she was part of the team that got the restaurant to number one on the San Pellegrino World's 50 Best List. In 2017, Jane was recruited to run the beverage program at Attica, widely considered Australia's best restaurant. In September of 2018, Jane passed the prestigious Master Sommelier exam, becoming the only woman in Australia to do so, and one of only 34 women in the world (and 274 people total since the 1960s). Jane has finally put her literature degree to use in publishing her first book.

Jane lives in Melbourne, Australia, with her husband, Jon, and her cat, Botrytis.

ABOUT THE AUTHOR

I heard "no" a lot of times when writing this book. It was a bold concept that required involved execution, and many people didn't want to touch it. So this is a big thank you to those people who said "yes" and continue to support me in this project and in life.

302

First, to Jane Willson at Hardie Grant for taking a chance on this book and dedicating so many resources to it. Her friendship and belief in me has been invaluable. To Emily Hart, who had to coax a slew of illustrations, design, text, interviews, and scholarship into a digestible form, and did so with positivity and precision. To Nadine Davidoff for her editorial expertise, which considered not only the words on the page, but the emotion and education behind it. And to my brilliant designer Lucy Sykes-Thompson, this book is only as powerful as it is because of your work.

Robin Cowcher, whose illustrations lit up the pages, has my eternal gratitude. Her role ended up being larger than either of us imagined, but she never faltered. I am super proud that my words are expressed through her images.

A special thanks to Francis Percival, Bronwen Percival, and Colin Spoelman for trusting their work in my hands. What they are contributing to their respective fields fascinates and inspires me.

Thanks to Stephen Barbara at InkWell for giving me the push to write a book (which can seem like a daunting endeavor). A big thank you to Emily Parkinson, whose original illustrations made this whole thing feel real. And to Rebecca Palkovics, who has taken on my interests and image like it were her own.

Thank you to all the sommeliers who gave their energy and insight to this book. An extra thank you to Kelli White, Rebecca Flynn, and Jordan Salcito for backing me up on a few chapters. And to The Violet Hour bartenders and Nashville songsters, thanks for the community and contributions.

Wine knowledge is created through a mosaic of sources: websites, books, magazines, articles, visits, and conversations. So a big thank you to those out there seeking to further the knowledge of our industry, who have and continue to make their imprint on me.

I could not have written this book without the culture, support, and understanding of Ben Shewry, Kevin McSteen, and the team at Attica. Thanks to them for putting up with me in a year where I prepared for the Master Sommelier examination, planned a wedding, and wrote this book. I'm sure they're as glad it is over as I am!

And finally, to my family and Jon. The dual hazards of ambition and anxiety have caused me to be quite inward-looking in my life. Mom, Dad, Beth, Randy, and Penny have shown me, without ever telling me, the value of looking outside myself. And they've done this while always encouraging my pursuits and massaging my psyche: no small feat. Jon has been my fact-checker, content-realizer, errand-runner, idea-bouncer, live-in cook, and love of my life through this whole thing. The words on the page are mine, but the spirit is his.

303

ACKNOWLEDGEMENTS

Published in 2019 by Hardie Grant Books,
an imprint of Hardie Grant Publishing

Hardie Grant Books (Melbourne)
Building 1, 658 Church Street
Richmond, Victoria 3121

Hardie Grant Books (London)
5th & 6th Floors
52–54 Southwark Street
London SE1 1UN

hardiegrantbooks.com

"Aussie Wine" chapter first published on GuildSomm.com as "Discovering the Wines of Australia," 2017.
Chart on page 37 inspired by "Whisky Flavor Profiles" by Christopher Ingraham; original concept and code by Kevin Schaul.
Chart on pages 114–115 based on chart from Colin Spoelman and David Haskell, *The Kings County Distillery Guide to Urban Moonshining: How to Make and Drink Whiskey*, Abrams, 2013; chart designed by Heesand Lee, Sebit Min.
Chart on page 144 with the help of Bronwen and Francis Percival, *Reinventing the Wheel: Milk, Microbes and the Fight for Real Cheese*, University of California Press/Bloomsbury, 2017.

A catalogue record for this book is available from the National Library of Australia

Vignette
ISBN 978 1 74379 532 3

10 9 8 7 6 5 4 3 2 1

Publishing Director: Jane Willson
Project Editor: Emily Hart
Editor: Nadine Davidoff
Design Manager: Jessica Lowe
Designer: Studio Polka
Production Manager: Todd Rechner
Production Coordinator: Mietta Yans

Colour reproduction by Splitting Image Colour Studio
Printed in China by Leo Paper Products LTD.